THE

TECHNOLOGY EDGE

OPPORTUNITIES FOR AMERICA IN WORLD COMPETITION

BY

GERARD K. O'NEILL

SIMON AND SCHUSTER   NEW YORK

For
Tasha

Copyright © 1983 by Gerard K. O'Neill
All rights reserved
including the right of reproduction
in whole or in part in any form
Published by Simon and Schuster
A Division of Simon & Schuster, Inc.
Simon & Schuster Building
Rockefeller Center
1230 Avenue of the Americas
New York, New York 10020
SIMON AND SCHUSTER and colophon are registered trademarks of
Simon & Schuster, Inc.
Designed by Karolina Harris
Manufactured in the United States of America

1 3 5 7 9 10 8 6 4 2

Library of Congress Cataloging in Publication Data

O'Neill, Gerard K.
The technology edge.

Bibliography: p.
Includes index.
1.High technology industries—United States.   2.High technology
industries.   I.Title.
HC110.T4048   1983   338.4'76213817'0973   83-20019
ISBN 0-671-44766-1

CONTENTS

ACKNOWLEDGMENTS

I acknowledge with appreciation the time, the care and the many insights contributed by those whom I interviewed for this book, all of whom are identified in the text. And I thank them for spending the further time to check draft manuscript sections based on their comments. Any errors that remain are my own. All the interviewees, in Japan, Europe and America, are people whose time is of great value to themselves and their companies. I appreciate all the more that they gave of it so freely. For his great assistance in arranging, participating in and following up on the interviews in Japan, I thank Mr. Kyoshi Asano. The plan for this book grew out of a series of conversations with John Brockman, and it is a pleasure to acknowledge his help in converting my deep concern with America's near future into a focused program of study and analysis.

This book concentrates on two closely related questions: What are the major new market opportunities for the next ten years, and how best can the United States use those opportunities to recoup its recent losses in worldwide economic competition? To answer those questions, my research necessarily explored business-management practices in several countries. I acknowledge with appreciation the several periodicals, named in the "Further Reading" section, from which some items of supplementary factual data were taken regarding specific successful companies. While this book is primarily based on interviews with chief executives, several fine books have recently appeared based on business-school analytical methods and on case-study analyses. Among them I would like to praise especially R. T. Pascale and A. G. Athos's *The Art of Japanese*

Management and T. J. Peters and R. H. Waterman's *In Search of Excellence*. In the limited areas where those works and this book cover similar topics, I am glad to find that there is complete agreement in our conclusions, different as were our working methods. I also acknowledge with humility that I am not an "expert" in the sense of being a business school graduate. In defense, I will only note that while much good work has been and is being done in our business schools, our toughest competitors right now are the Japanese, and they have no business schools at all.

In the course of research it was my good fortune to meet and interview a number of extraordinary individuals working to realize the new large-scale industrial opportunities that form this book's central theme. I owe them special thanks, because their work proves that the few remaining years of this century will be, for those nations which meet the challenge, a time of unprecedented opportunity.

GERARD K. O'NEILL
Princeton
May 1983

AUTHOR'S PREFACE

One retains images, some profound, others simply funny, from an interview. When I was escorted into the top-floor suite at SONY headquarters to meet Masaro Ibuka, a founder of the corporation, I was startled by three almost identical, very pretty secretaries, all in a line, all dressed in uniform blazer and skirt, bowing in unison.

The first of the images from a meeting with Konosuke Matsushita, founder of the $12-billion-per-year firm whose products carry such names as Panasonic, Quasar, Technics and National, was a reminder of Japan's high population density. My trip from Tokyo to Osaka to see Matsushita was on a JAL Boeing-747. The plane was configured all-economy with 450 passengers, and every seat was occupied. But the long-range Boeing was being used for a flight of only forty-five minutes. Then, recalling my visit to Matsushita's Central Research Laboratories, I think of the deep carpets, wood paneling and works of art in the presidential wing, and a lunch there at an enormous table set for just two places, my own and that of Councilor A. Nagano.

Matsushita, at 87, inspired the respect due to someone with a strongly spiritual quality. He walked slowly into the quiet interview room. As a thoughtful and courteous gesture, he was carrying a copy of the Japanese edition of a book of mine, published by his PHP Press. I recall, from the interview that followed, his words, spoken softly yet with clarity and conviction: that the view of humans as essentially weak and criminal is wrong. Humans will continue to make progress. The degree of control they exercise over the Earth will increase. They have the responsibility to be the saviors of all other creatures . . . and because human beings pursue ideals, eventually those ideals will come true.

On the time scale of centuries, Matsushita saw the present economic conflicts between Japan, the United States and Europe as "ripples on the sea. During competition in adolescence, some people will win, others lose, some get sick or injured; but in the long run, all will become adult." Under Konosuke Matsushita's direction, his company has been extremely successful in competition with American and European firms, and I was impressed by how much its success had been enhanced by his strongly philosophical sense of responsibility to society. Later, in the United States, I found striking American parallels to the Matsushita history, but nowhere a match to the tranquil, uniquely Japanese image of the room where we met.

GERARD K. O'NEILL
Princeton
May 1983

INTRODUCTION

This book is about big new market opportunities—not a million dollars or even a billion, but tens and hundreds of billions. Those opportunities are available to the nations of the world in the decade ahead, and I am concerned that the United States may let them go by. If it does, our country will begin that long slide downward which so many other nations have taken—and from which there is no returning.

My concern is so deep that I left a secure academic life to help, as much as one man can, in establishing a new company. That company, the Geostar Corporation, will be discussed in later chapters. It plays a small but essential part in opening one of the major new market opportunities for America.

This is not another "Japan book," but Japan is our most formidable industrial competitor, so we will begin with a searching look at Japanese industry in Part I. Nor is this a book about management —but we would be foolish not to heed the lessons we can learn by analyzing Japanese methods for success. It is no accident that Japan has seized upon science and technology as its most effective weapons. New technological opportunities have been at the core of every case history in which American industrial leadership has been toppled or shaken.

In Part II we will explore the six major technological opportunities of our time. They carry no flag, and no country may claim them without a major continuing effort. Three of the six have already led us into battle on a world scale. The other three opportunities are as yet virtually unrecognized. Our neglect of those opportunities would complete American defeat; our vigorous exploitation of them could secure a healthy and peaceful victory.

In Part III we will investigate uniquely American developments that are already remarkably successful and productive—and very difficult to imitate. We will find that the most successful American corporations have been using, for many decades, techniques that are now being called "Japanese."

My approach is factual rather than theoretical. In my search for a course for our nation out of the severe problems it faces in the closing years of this century, I used the hands-on, pragmatic method that experimental scientists have learned to trust. To find what is really happening worldwide in our era of explosive technological change, I interviewed more than forty people in Japan, North America and Europe. All—company presidents, directors of research, founders of the most successful major corporations—had shown their ability to pick the most productive technologies and business methods. From those interviews I came away astonished by how far certain critical developments had already gone, in most cases far beyond the points generally realized by the public. The quotations that follow will give you the flavor of those interviews.

In the past four years of genetic engineering research we accomplished what we had thought would be a twenty-year program . . . each problem we'd imagined to be a serious barrier melted away when we attacked it. And every time, we were struck by how mechanical the genetic structure was. It turned out to be a system built out of precise shapes and sizes, fitting together like a child's Erector Set. I don't like to be overly dramatic, but I can see a time not far off when disease will be conquered to such an extent that the main cause of death will be accidents.

—Dr. Ronald Cape, Chairman of the Board, Cetus Corporation

My firm did the incorporation papers for a high-tech Silicon Valley start-up firm less than two years ago. We took fourteen thousand shares for fourteen hundred dollars—ten cents a share. Thirty days later, major financial interests put four and a half million into it at four-fifty a share. The next round was at sixteen dollars per share, and it's split twice since. The new round will be at ten dollars— equal to forty before the splits. So it's ten cents to forty dollars in eighteen months: four hundred to one.

—Mario Rosati, partner in Wilson, Sonsini, Goodrich & Rosati

Our business expanded by over fifty percent per year from 1975 to 1980, and we're earning a profit of about thirty-three percent each year. In our factory at the base of Mount Fuji we can produce five days a week, three shifts every day. But of course, there are human workers present only on the five day shifts out of the fifteen. The factory is operated entirely by robots directed by a central computer system for ten out of the fifteen shifts.

          —Dr. Engineer Seiuemon Inaba, President, Fanuc, Ltd.

We'll put the M-301 on the market in 1986, as a six-passenger airplane for the charter, business, family and personal markets. It will cruise at twenty-five thousand feet, and according to the wind-tunnel tests it will deliver a highway equivalent of three hundred forty miles per hour, while getting a mileage of twenty-one miles per gallon. We're building it to be stronger than an airliner.

          —Roy Lopresti, Director of Research, Mooney Aircraft Company

My immediate concern is the economic survival of our nation in a time of unprecedented challenge from abroad, made possible by technological change occurring at an ever-increasing pace. But I have a longer-term concern over issues that transcend nationalism. That concern is visible in the choices I make of potential major new technologies to explore. To gain my interest and support, they must satisfy three conditions on a global scale: they must serve individual human beings better than the older technologies they replace; they must be more efficient in their use of energy; and they must be more benign to the planetary environment.

I set those conditions not only because it is our human responsibility to do so, but also because they are exactly the conditions which must be met by any technology that can generate new wealth rapidly in a free-market economy and arouse no citizen resistance. Japan is exploiting just that class of technologies in its drive for economic domination.

As an American, I cannot help being angry—not at the Japanese for succeeding, but at the forces of timidity, shortsightedness, greed, laziness and misdirection here in America that have mired us down so badly in recent years, sapped our strength and kept us from equal achievements.

As we will see, opportunities exist now for the opening of whole new industries that can become even greater than those we have lost to the Japanese. Are we to delay and lose those too? Perhaps;

but this book is above all a call to arms, an appeal to Americans to straighten up and perform as we have never before had to perform except in war—and to maintain that performance not just for a short time but steadily, over years. Can we do it? Certainly. Will we? That is far more doubtful; but on that issue I leave the last word to a Japanese.

He is Rokuheita Numata, Senior Editor of Shincho-Sha, publishers of the Japanese edition of this book. Near the close of World War II, Numata, not yet 20, was a sailor in the Japanese Navy. In the same years, I was a young volunteer in the U.S. Navy. When we met, many years later, I found that he retained a nostalgic enthusiasm for the beautiful fighter planes of that era—especially American planes, the P-38 and the P-51. At dinner one night in Tokyo, Numata tried to guess the final outcome of the bloodless war in which our countries are now engaged:

You know that the question of going to war with the United States was argued fiercely between the Japanese Army and Navy. General Togo and his Army clique pointed to America's softness, to its divisive racial problems, to the constant squabbling of its politicians, and were contemptuous. One hard push, they argued, and America would topple. But many of the admirals had gone to college in the United States, and knew its people better. They could answer Togo only with feelings, not with facts. And they said: "Don't awaken the sleeping giant!" The generals brushed that aside, and ordered the attack on Pearl Harbor. We all know how successful that war looked for Japan in the beginning; but we also know how it ended. I have a feeling that the sleeping giant is about to awaken again.

ONE

A SEARCHING LOOK AT OUR STRONGEST COMPETITOR

CHAPTER 1

JAPAN CHALLENGES THE WORLD

After 1868, when Commodore Perry's U.S. Navy warships steamed into Yokohama harbor, Japan's physical isolation ended forever. There was an ancient proverb (spoken as *"Wa-kon kan-sai"*) which meant wisdom from China, applied with Japanese spirit or soul. It changed. As Japan raced to catch up with America and Europe in the following decades, the proverb became *"Wa-kon yo-sai"*: Western ideas, but interpreted, as China's had been before them, in a uniquely Japanese way. The proverb has changed yet again. Its new form is *"Wa-kon ka-ghi-sai"*: the wisdom of science and high technology, but filtered through the mesh of Japanese values and patterns of thinking.

Americans, who coined the phrase "know-how" early in this century, are finding the unexpected turns of our national fortune threatening. A loss of market share forced the firing of a third of the workers in America's automobile industry. The U.S. consumer-electronics industry has been largely wiped out by Japanese-import competition. If present trends continue, the United States will fall to second-class rank among the leading industrial nations by the early 1990s. That loss affects Americans of every political persuasion, for the production of wealth is a basic need of society. Conservatives whose main concern is Soviet aggression have reason to fear that America's loss of economic strength may tempt the U.S.S.R. into military power plays closer to us than Afghanistan. Liberals more concerned with social issues can already see the economic war taking its toll as the lowest socioeconomic classes lose their jobs and turn to social programs that are increasingly overburdened and unable to support them.

It takes intelligence and literacy to use high technology effectively, and Americans are at a disadvantage compared with the Japanese. A specialist in comparing intelligence between cultures, Dr. Richard Lynn of the University of Ulster, has found that the average Japanese scores a full 10 percent higher in IQ than the average American. Close to 40 million Americans have IQ levels below 85; by contrast, only 5 percent of the Japanese population scores that low. Poor schools and inner-city problems keep millions of our children from achieving the intelligence they are capable of, and that our nation desperately needs. Fifteen percent of our population is functionally illiterate. Yet fewer than 1 percent of the Japanese are unable to read, although Japanese ideographic writing is far more difficult to learn than an alphabetic language like English.

Our industries also lose to foreign competition because we persist in static thinking while the reality is constant change. In the world of fast-evolving technologies, growth rates and reinvestment are more significant than the current size of a company or its dividends to stockholders. It is easy to ignore, as many companies oriented only to quarterly profits have ignored, the threat posed by a small competitor with a high growth rate. Growth of 20 percent per year is modest by the standards of high-technology industries, but even that rate allows a rival to grow to 1.44 times its original size in just two years. In just thirteen years it can grow to ten times its original size. That kind of high growth, over a period of hardly more than a decade, brought Hitachi, Matsushita and Toshiba, all names barely known in the United States in 1967, to the ranks of the top forty corporations in the world by 1980.

The personal computer, originally very much an American preserve, illustrates how rapidly a market can be lost to such fast-growing competition. In the late 1970s, Apple Computer and Radio Shack, the dominant manufacturers, accounted for half of all production worldwide. In 1979, Japanese computer manufacturers were able to hold only 20 percent of their own domestic market against them. But just one year later, the U.S. share of the Japanese personal-computer market had dropped from 80 percent to 20 percent; and in the following year, seven leading Japanese electronics manufacturers, NEC (Nippon Electric Company), Canon, Hitachi, Oki Electric, Sharp, Toshiba and Matsushita, entered the U.S. domestic market. By 1982, they accounted for nearly 40 percent of the more than $2 billion in personal-computer sales within the United States. And that is something of an underestimate, because a great deal of American-labeled computer peripheral equipment is made in Japan. In 1981, IBM developed its own personal computer on a crash basis, to grab a share of that market. The "IBM" printer it sold to go with its personal computer was made by Seiko, and was identical to a model sold by Seiko under the "Epson" brand name. The Epson printers were also sold by Xerox and other U.S. manufacturers under their own labels.

"LIMITS TO GROWTH"

There is no single reason for U.S. defeats in the war for international markets, but a philosophy sometimes referred to as "limits to growth" that was fashionable in the 1970s may have contributed to those failures through its effect on governmental policies. Administrators and members of Congress, who looked toward a static society as a desirable goal, failed to provide the spurs for investment in plant modernization and in the opening of new markets based on emerging technologies. The business-management philosophy popular in the United States during the late 1960s and 1970s also contributed to later American industrial defeats. Industries were run by cost accountants with little interest in technology and even less understanding of it. They were infected by a kind of merger mania. Akio Etori, editor of *Saiensu*, and Professor Gene Gregory of Sophia University in Tokyo, in a 1981 article titled "Japanese Technology Today" (JTT), said:

> Managers of [U.S.] manufacturing enterprises are more concerned with managing assets than with managing people and technology. Management tends to be impersonal; people and technology are expendables, traded in the marketplace of corporate assets by wizard-conglomerators. Japanese manufacturing firms tend to be managed by engineers who have had a lifetime of experience in working together with colleagues in the same enterprise. As a result Japanese companies are likely to be more responsive to changes in technology.

PATIENT MONEY

With strong technical orientation, and "patient" money from banks willing to make long-term low-interest loans, Japanese companies invest far more heavily in new plant equipment and in research and development (R&D) than do their U.S. competitors. During the 1960s and 1970s, private investment averaged 35 percent in Japan, compared with 25 percent in West Germany and 17 percent in the United States. In the high-tech semiconductor business, Japan's manufacturers plowed back even higher percentages of their capital into building for the future. Over the five-year period centered on 1970, they put 60 percent of their sales income into building new plants, and another 25 percent into R&D. And each time a new technological wave swept over the semiconductor industry, the Japanese manufacturers renewed their commitment: when Very Large Scale Integrated (VLSI) circuits were being developed about 1980, several Japanese firms reinvested more than 21 percent of sales revenues in new R&D and another 22 percent in new production facilities. Even after the industry matured, they were willing to accept small profit margins, like 1.2 percent for NEC, the largest Japanese maker of integrated circuits, and 2.6 percent for Fujitsu. In Japan, stockholders' equities are highly leveraged, so those small profits are quite satisfactory to Japanese investors. The JTT article also said:

> Since industrial structural change is likely to be more rapid in Japan than in other advanced countries, the focal point of innovation and investment will continue to shift to Japan—especially in key technologies such as electronics. While R&D investment in electronics in other countries has been declining, it has been rising steadily in Japan.

QUALITY CONTROL

Poor quality control in American products is responsible for loss of competitiveness. A director of the SONY Corporation spoke of a product called "SOS"—"semiconductor-on-sapphire"—a material that manages to conduct heat relatively well, but insulates against the flow of electric current. It can greatly increase the power output available from an integrated circuit, so companies in the United States and Japan scrambled to produce it. The SONY director explained that

Making SOS is a delicate process, because it must be made at a high temperature and then cooled very slowly. The quality coming out of the American plants was quite poor—the chips cracked; and when a control engineer checked it out, he found the defective chips were those made near the end of the working day. The workers cooled down the furnaces too fast in order to be first out to the parking lot at five P.M. In Japan, the workers just stayed until the job was done right.

Toshio Numakura, General Manager of the Hitachi Institute of Technology, described Japanese workers as "nervous about the reputation of their company. They take quality control very seriously because of that sense of identity and responsibility." The Nissan Motor Company plant in Kamakura, where Datsuns are produced, many for export to the United States, exemplified this. My tour of the Nissan plant was conducted, like so many in Japan, with regard for ceremony. I was ushered into a reception room decorated in Western style with couches and a low table. It carried miniature Japanese and American flags. The Deputy General Manager, Kenichi Shima, and the Engineering Production Manager, Nobuki Yamamoto, were waiting, dressed in the same style of blue uniform coverall that I was to see on every Nissan employee. Shima's comment on quality control was made without hesitation: "Our high quality comes from the heart of the worker."

In a quite different industry, Nitto-Boseki is described by an American expert as the most advanced fiber-glass-manufacturing company in the world. Its President, Dr. Kesaharu Kasuga, said:

For the Japanese electronics industry, we manufacture zero-defect yarn and cloth. For large composite components of high strength like automobile springs, we manage to get even the smallest detectable inaccuracy down to a very low level. We use the Haguruma principle [the "matching of gears"—the identification of one man with one responsibility]. The biggest difference between our quality and that of U.S. manufacturers is in making the glass fibers themselves. We use a very large number of holes for extruding the molten glass, and as the fibers emerge we cool them slowly, with air, rather than quenching them very fast with water sprays. That way we get a very even tension, higher strength and more uniformity.

The message was clear. The JTT article also argued that America could never hope to equal the Japanese level of human responsibility in quality control:

> Our Japanese mode of management gives better quality because it derives from the absence of those dichotomous adversary relationships which pit management and workers against each other in eternal conflict in Western enterprise.

But there are limits to the human control of quality, and can Americans reach them? The astonishing high level of reliability achieved during the Apollo lunar flights was reached entirely through a zero-defect policy carried out by American workers and management working in partnership. At the Honda Motor Company, Chairman of the Board Hideo Sugiura had firsthand evidence that individual American workers could work as effectively and as carefully as their Japanese counterparts. Quality control as a specific discipline was pioneered in World War II by an American engineer, Dr. W. Edwards Deming. Every year since 1950, the Japanese Union of Scientists and Engineers has awarded Deming Prizes to industries, and there is fierce competition for those prizes. But, said Sugiura:

> Quality-control techniques . . . as such can't be exported, though management philosophy can be. We know what does work in the United States, because for several years we've had a plant near Columbus, Ohio. At first we couldn't get quality, because there was a U.S.-type stratification of workers/supervisors/managers. So we sent over about thirty Japanese as managers and supervisors, to break the stratification and achieve our kind of unity. Now the productivity and quality coming out of the Ohio plant are at least as good as anything out of Japan—maybe even better. The kind of small-group unity and group quality control that are usually called "Japanese" sprang up spontaneously in the Ohio plant, out of the workers' acceptance of the philosophy. It couldn't be forced. I recall one incident when I dropped into the plant unannounced after hours one night. I found one "quality-control circle" of American workers still working, on their own time. When I asked, they told me they'd spotted some paint defects on the production line and had decided to stay and make them good before leaving. No one would have been surprised at that in a Japanese plant; but this was Ohio.

Sugiura pointed out that most of the Japanese managerial staff had been replaced by Americans as of 1982. Only five Japanese were left, but they filled the top three positions at the plant and two of the four at the next level down.

Both at SONY and at TDK Electronics, new limits are being reached in quality control. Those limits have a political as well as a technical component, because Japanese firms are finding it necessary to set up manufacturing plants in other nations to avoid tariff barriers. At SONY, production engineers were convinced that they could eliminate the problems of quality control in VLSI-chip production only by going to the workerless factory. VLSI chips have circuitry so microscopic that a grain of dust landing on one can be ruinous; until they are encapsulated, they are extremely vulnerable. Every chip that must be rejected raises the break-even price that must be charged for the rest, in a fiercely competitive market. SONY engineers theorize that in a factory with no human respiration and none of the hair and skin cells humans continuously shed, they could achieve near-perfect control over dust, and very low reject ratios on VLSI-chip production. In their view, they will have to reach that level of total automation within this decade.

TDK Electronics is unusual as Japanese firms go, because most of its sales are in a narrow range of products: recording-tape cassettes and ceramic capacitors. The design of audiocassettes that has become an industry standard was first developed and introduced by Phillips of Eindhoven in the 1960s. TDK manufactures those cassettes under license, and dominates now as a result of its outstanding quality control. Many of the audio- and videocassettes that bear the names of other manufacturers are in fact produced by TDK.

The mastermind of TDK's quality-control program is Sho Masujima, TDK's Executive Managing Director. To reach TDK's goal of zero defects, he had restructured quality-control methods and production processes so that zero-defect products could be manufactured even when human machine operators made some errors. In ceramic capacitors, performance depends on the exact details of a witch's brew of barium, titanium and other chemical elements. TDK extends its quality control well upstream in the production line, to the ratios of area to volume for the barium and titanium components, and to their crystal structures. Those ratios measure particle size. Once they are closely controlled, the product is so superior that it can tolerate human errors such as getting the mix too hot. With the help of a scanning electron microscope, TDK researchers carried out exhaustive studies of grain size, shrinkage, loss of binder and other parameters all as functions of temperature and time. From those studies Masujima was able to establish a range for each variable downstream that could be affected by human error, and when the ranges were set correctly, the entire process became much more tolerant of human mistakes. Masujima said:

> The zero-defect quality analysis is very expensive, but once it is completed it pays for itself quickly in two ways: first by the money saved through much lower reject ratios, and second by a rapidly increasing market share. One of the best things about it from our viewpoint is that it works in all countries, even those where the educational level or the motivation of the work force is relatively low.

Masujima was far from secretive about his methods—enthusiastic, in fact, about exporting them to the United States. But he doubted that U.S. firms, other than those wholly owned by Japanese, would adopt them:

> The problem with U.S. companies is their short-term expectations. It would take a fundamental change in financial-management policies, away from the emphasis on short-term profits, before American firms would adopt my method. But I wish they would; I'd like to see the zero-defect method used worldwide.

It is difficult for TDK to get exact numbers on defective cassettes, because the customer is the one who finds them, and he doesn't, as a rule, report back. But capacitors are sold mainly to original-equipment manufacturers, who keep their own records and report to TDK monthly. By 1973, TDK's partially automated assembly lines, together with strict quality controls, had brought the company's defect rate on capacitors down to about 10 in a million capacitors sold. That didn't satisfy Masujima, so to achieve a still higher level of quality he introduced the upstream-measurement approach. The results were impressive: by 1978, the production-defect rate was down to 1 per million, and now, after still more automated machines have been introduced, only about 1 in 5 million capacitors shipped is defective. Masujima is working toward still higher quality, with a goal of not over 1 faulty capacitor out of 100 million.

PRODUCTIVITY

If one side of the coin of competition is quality control, the other is certainly productivity. Many of the U.S. failures are in just those high-growth industries where failure can be least afforded, because the potential for earnings is so great.

The case of the vanishing U.S. printer industry is an example. American firms pioneered the relatively inexpensive dot-matrix printers, machines that could produce any letter or number by combinations of a few, usually seven or nine, tiny needles magnetically driven to strike an inked ribbon over paper. In 1979, U.S. firms accounted for nearly 100 percent of the sales of dot-matrix printers. The market for them exploded then because they were affordable as add-ons to personal computers. But the American firms were unable to expand their production and control quality sufficiently to satisfy the demand, so their delivery delays gave firms in Japan like Ricoh, Okidata, Seiko and Itoh the opening they needed. Less than eighteen months later, those Japanese firms were selling half the dot-matrix printers bought in the United States, and six months after that, their share of U.S. sales had risen to 75 percent. Japanese dominance of that U.S. market is estimated to approach 100 percent by 1985, and printer sales then will be hitting $1 billion per year while still growing very rapidly.

As the 1980s began, the American electronics industry as a whole and within it the computer industry were still larger than their counterparts in Japan. But as the JTT article says:

> As computer production moves from a narrow industrial market to a mass-market with global dimensions, the Japanese computer industry will close the gap with the U.S. industry by the end of the decade, erasing the last major advantage left to American electronic equipment manufacturers.

MILITARY AND MASS MARKETS

The Japanese argue that many U.S. electronics firms have become stodgy and uncompetitive because their sales are to military customers. Indeed, military buying decisions are usually made with input from only a very few people, and the political and bureau-

25

cratic reasons for them often outweigh the technical. The JTT article says:

Although the American electronics industry is considerably larger than the Japanese industry, as much as 45% of its production is geared to special military requirements. The Japanese industry, on the other hand, is totally market-oriented, offering a far wider range of products than the more narrow-based U.S. industry. This . . . provides the Japanese industry with extraordinary flexibility in the marketplace and enables it to sustain high R&D expenditures.

Dr. Michiyuki Uenohara, Senior Vice-President of NEC, makes the same point: "Japan has developed its industries to cater to the needs of consumers. By contrast, European countries and the United States have been able to do a good deal of business because of military . . . procurement markets."

The criticisms seem well based, especially when one considers that military buying decisions often take years, while the decision of an individual consumer to make a purchase can be made almost on impulse, without committee meetings or reports. One large American company, Hewlett-Packard, maintains a policy of neither seeking nor accepting military contracts.

PRODUCTION ENGINEERING

Production engineering, the second key to Japan's fantastic success in taking growth markets away from the United States and Europe, manifests itself directly through increases in productivity. As the 1980s began, the nationwide average productivity in the United States, measured in terms of output per labor-hour, was still slightly above that of other advanced nations. But the situation would soon reverse, because the growth rate in American productivity was so much slower. In an earlier period, from 1948 to 1965, U.S. productivity had grown at a fairly healthy rate of 3 percent per year. But after that it declined steadily, averaging a weak 0.8 percent in the mid-1970s, to − 1.0 percent by 1979. In the same period, 1968–1978, Japan's productivity growth rate averaged 7 percent. Since then it has gone higher still for certain outstanding Japanese corporations. Mazda Motors, for example, averaged a 15-percent-per-year productivity growth rate going into the 1980s, doubling in five years its output per employee. The first oil crisis of

1973–74 has been blamed for some of the U.S. woes in productivity, but that is a lame excuse; Japan's dependence on foreign oil is total, ours only partial.

A fundamental strategic mistake was made by U.S. industry in the late 1960s and 1970s. It is just the kind of mistake one would expect of money managers ignorant of technology, indifferent to building strong management-worker cooperation and excessively focused on short-range profits. Many U.S. industries in that period moved labor-intensive operations to cheap-labor areas in Taiwan, Singapore, Malaysia, Hong Kong and the Philippines. In contrast, Japanese industries invested in expensive, automated production equipment at home. That is paying off now in high productivity, high reliability and low costs, while the U.S. firms helplessly watch labor costs rise rapidly in all of Asia except for mainland China. The JTT article sees Japan's concentration on increasing productivity as a manifestation of Japanese industrial philosophy: "Unlike Western enterprise, the primary purpose of which is seen as profit maximization, the Japanese enterprise has as its principal objective the creation of wealth, adding value through production."

Production engineering has two components, and Japan has led in both. One is the detailed design of new products in a short time, while making them effective, attractive, elegant in their operation and easy to produce. That concerns the engineer at his drawing board or, more recently, at his computer-aided-design (CAD) console. Japanese corporations are willing to invest far more in such detailed design than their U.S. counterparts, and they can afford to do so because engineers are more plentiful in Japan than in the United States—per capita, Japan has about three times as many. And although the number of engineers graduating at the bachelor's-degree level in the United States rose slightly during the 1970s, the number of American engineers obtaining Ph.D. degrees actually fell by 30 percent during the decade. A popular joke says, "When emission and gas-mileage requirements were slapped on the car industry, Japanese carmakers hired more engineers. U.S. carmakers hired more lawyers."

ENGINEERS
The Japanese have made the more profitable choice. Even the reduction in the U.S. output of Ph.D. engineers is a consequence of the overall engineer shortage, because it shows that fresh Bachelor

of Science graduates are being pulled into industries desperate to hire them, rather than taking the extra years to get the higher degrees. The shortage is damaging future prospects for rectifying the problem, because it is also decimating the ranks of the engineering faculties at American universities. Ph.D.-level professors are being lured out of their teaching posts into higher-paying industrial jobs. American industry can't be blamed for "eating its seed corn" in that way—it has no choice, given the supply and demand in the market for engineers—but the fact remains that in 1980, from 10 to 15 percent of the faculty positions in America's engineering schools were vacant. Starting the 1980s in that weak position, the United States had no way, without some innovative teaching methods, to rectify an imbalance that had Japan, with half the population of the United States, graduating about one third more electrical engineers than we did every year. The United States in the 1970s, and now even more in the 1980s, is caught in a vicious circle: too few engineers, so salaries in engineering are forced so high that industry can't afford to assign enough engineering time to the development of a new product. As a result, that product is underengineered and relatively crude, and can't compete in the international market. We lose still another high-growth export market, the Japanese gain it and the U.S. economy takes another dive. That puts still more pressure on U.S. industries to cut corners for economy, and leaves still less money to be given to U.S. universities to improve their salaries for engineering professors. Hisashi Shinto, President of Nippon Telegraph and Telephone, sees two ways in which the U.S. shortage of engineers has hurt us:

> In 1955, I saw many capable young engineers working in U.S. factories to improve manufacturing technology. And they trained the workers to perform better. But today I cannot find such qualified people . . . working on the shop floor. Most production managers now are not college-educated. They do not have the intellectual capability to devise improvements in manufacturing. That is the key reason why the productivity gap [between Japan and the United States] is growing.

Shinto also notes that U.S. firms, always short of engineers, cannot make the small modifications in their standard products that would make them competitive in export sales.

CONSUMER ELECTRONICS

The sheer numbers of engineers in the United States and Japan are only part of the U.S. problem. As William Meserve of the A. D. Little Company observed:

In America during the 1960s and 1970s, consumer electronics was viewed as a stodgy, lowbrow, low-growth business, and the brightest young American engineers went into computer firms and Silicon Valley companies. But in Japan, most of the brightest young engineers went into consumer electronics in those years.

Those young Japanese engineers of the 1960s and 1970s are now experienced project supervisors in the prime of their mid-careers. Dr. Uenohara of NEC made a similar comment about the attraction of the high-cost big-hardware projects over consumer electronics for young engineers in the United States: "For technicians and engineers, [those areas] are far more exciting to work in than those catering to the masses' needs."

JAPANESE MANAGEMENT

We saw, in the case of Honda's introduction of a Japanese management philosophy into its Ohio plant, that the Japanese approach works very well, even with American workers from a distinctly non-Japanese cultural background. It should, because it is human and universal rather than nationalistic. The philosophy involves a respect for individuals and the involvement of everyone in the production process, from managers to production-line workers, in making that process work better. The "Japanese approach" also works with engineers, and engineers are going to become even more important in American industries as productivity increases require increasing levels of automation. L. K. O'Leary, Assistant Vice-President at AT&T, writing in the *New Jersey Bell Journal*, points out that some American companies have been using their own versions of the "Japanese approach" for a long time. As he says, "In a few businesses, notably the go-go electronics industry, whole companies started small-and-human and were smart enough to stay human after they got big, or even huge." O'Leary believes the new generation of American workers

isn't asking to be made happy, or to be given something. Most of them are asking to be a part of the decisions that affect their futures, to be consulted and listened to in matters concerning themselves, and want to find some meaning in the big chunk of their lives they spend at work.

In his view, many American business managers can't deal effectively with this reality because

our business cultural base follows a military style, with leaders and followers. In the American corporation, we have lonely commanders giving orders in a hierarchy of power where the lesser beings respond to directives from above. When introduced to the New Worker problem, such a manager's first reaction is denial that it exists, followed by denial that it is relevant to results, followed by denial that there is a solution.

O'Leary recommends that corporations

change your whole management selection and promotional systems, if these systems were set up to identify and promote the authoritarian, results-at-all-costs manager we can ill afford. Change your entire reward system, if it pays off for short-term results and ignores their long-term human and economic costs. It means changing, in most cases, the basic culture of a business—and that takes years.

A general problem in American industry is the inability, in many cases, to hold together and strengthen engineering design groups that have proved their outstanding ability. Tracy Kidder gave an account of an intensive, successful design effort in *The Soul of a New Machine*. The case he studied was at Data General, a manufacturer of minicomputers. In 1979, Data General's design leader Tom West recruited an outstanding group of young electronics engineers, to do the nearly impossible task of designing, debugging and bringing to market in just one year a complex new 32-bit minicomputer. The machine became Data General's Eclipse MV-8000. The company's talented President, Edson de Castro, had defined the parameters for the new machine. To be a success, it would have to be compatible with all the programs written for Data General's earlier computers, as well as with new, far more complex programs. That made the task of designing it very difficult.

The engineers, under intense pressure, formed themselves into shifts and worked day and night for many months. They forged the bonds of knowledge, trust, respect and loyalty that made them a team, capable of far more than any one of them could have achieved on his own. As they worked, most of them assumed they would be kept together as a group and given another project; in their phrase, "another chance to play the game." Yet when they had achieved their goal, they were given no incentive to stay together as a group. Most of them left, and the teamwork they had built up, a priceless, though intangible, asset to the company, was allowed to dissolve. A Japanese corporation competing with Data General would not have permitted such a loss.

TIME FRAME

A nation tends to freeze its ideas at the time of its greatest successes, and once those ideas are set they are not easy to change. The period of least challenge to American "know-how" was the first sixty years of this century. The Great Depression notwithstanding, that was a time when American industry dominated world production. The important industries were "low-technology" by modern standards, and their products changed little from year to year: automobiles, steel, railroads, buildings, ships, coal, oil. That made it tempting for American companies to enlarge their profits by the easy route of establishing cartels, rather than the more difficult and challenging route of higher productivity and quality control.

Society's answer to the cartel was the antitrust system of laws set up by President Theodore Roosevelt and his successors. That system, vitally needed when it was established, froze American industry and American government into an adversary relationship from which they have never escaped. Limited communications and the absence of a modern high-speed interstate highway system made it difficult for American workers to escape unsatisfactory working conditions. Industrywide trade unions grew up—vitally needed at the time to protect American workers against exploitation by their employers. But those unions, now grown to great power, institutionalize an adversary relationship between industry and labor. Those two adversary relationships, industry/government and industry/labor, are heavy handicaps in competing with the Japanese. The circumstances we find ourselves in have changed, but our institutions have not.

Japan's ideas about production and competition have also been formed in a period of success—but for Japan that period is recent, from 1960 to the present. Those ideas, and the corresponding Japanese institutions, seem far better suited than ours to the modern world. The Japanese understand that they must export and compete internationally to survive; that market opportunities increasingly will involve high technology, with products evolving on a time scale not of decades but of one year or even less. The richest returns will be from products that have high value-added, and can be made without heavy use of energy or materials. The Japanese understand that in an era of permanent fierce competition, a company must enlist the enthusiastic cooperation of all its employees if it is to increase its market share. Those realities have led Japanese corporations to avoid merger mania and concentrate instead on a far higher degree of vertical integration than is common in America.

This high degree of vertical integration is illustrated dramatically in one of the most complex products in Japan, Fujitsu's FACOM M-382 mainframe computer. Everything in the machine, down to the extraordinarily complex multiple-layer circuit boards and the special VLSI chips mounted on them, was made by Fujitsu. And Fujitsu is typical rather than exceptional in its vertical integration. SONY, Pioneer, Toshiba, Sanyo, Sharp, Seiko and many other Japanese electronics companies have also invested heavily in order to develop their own in-house capability for making semiconductor chips, difficult and complex as that task was.

To generate the big cash flows necessary to sustain the substantial investments in research that are needed, Japanese companies have concentrated on the consumer market for their products. Unlike the slow-moving industrial and military markets, the consumer market can respond to an attractive new product as quickly as buyers can take out their credit cards. From that response, the highly vertically integrated Japanese companies can very quickly shift production facilities to the new product and can broaden it to a complete line. Those companies, with their in-house component facilities, aren't slowed down by the outside suppliers' delays that plague U.S. corporations. A few years ago, SONY illustrated that sort of response when it brought out its Walkman audiocassette player. According to a SONY director, the Walkman was a "sleeper" at SONY. No one in the corporate high command had suspected how popular it would be. But as soon as its meteoric rise in sales became clear, SONY was ready to churn out Walkmans by the millions, and within a year had expanded the line to five different models.

PROLIFERATION OF OUTLETS

Japan itself is the testing ground for the new products brought out by Japanese industries. The Japanese people, educated, affluent and receptive to high-technology products, are a living consumer-product-evaluation institution that pays Japanese industries for the privilege of trying out new products before the products—or rather, those which survive in competition—are exported to the rest of the world. One sees that evaluation process at fever pitch in the Aki-Habara district, Tokyo's retail electronics center.

Each store in Aki-Habara is huge, with several floors of display space devoted to electronics alone. And every floor, up to the fifth and sixth, has window displays fronting on the street. Inside a typical shop, whole floors are devoted to each major manufacturer. Matsushita, largest in consumer electronics, is represented with its Panasonic, Quasar, National and Technics lines of products. But its competitors like Toshiba, Aiwa, SONY, Hitachi, Sharp and the rest are in head-to-head competition with it, offering a product for every one in Matsushita's lines. One has the choice not just of a few video tape recorders, for example, but of forty or fifty, from a half-dozen major manufacturers. And where, in a shop in the United States, one might find ten or fifteen educational pocket calculators intended for children, in any of the shops in Aki-Habara one finds sixty or seventy. Those electronic products are textbook examples of high value-added to materials that cost very little. And those Japanese-made products absorb more than half of what was already, by 1980, more than a $4-billion-per-year Japanese output of semiconductors. They provide a nearly recessionproof buoyancy in the market: in 1981, Japan increased its total semiconductor sales by 24 percent, while America's were down by 2 percent and Europe's were off by 15 percent.

Japan's concentration on production engineering as a key to gaining market share and its concentration on consumer electronics as the hottest possible market have turned the entire nation into a kind of money machine, a cash generator. Japanese electronics and computer companies are generating so much cash that they can provide for 75 percent of their new-plant and research needs without going outside to banks. Some are even repaying bank debt. Yoshiaki Enomoto of Japan's Long-Term Credit Bank complains that he wants "to do more business with the electronics companies, but they are flush with money." Tax laws in Japan help make it easy for companies to provide for new growth out of their own liquidity, because those laws permit accelerated depreciation of new plant equipment, and reward heavy expenditures in R&D. In a typical year, the Ministry of International Trade and Industry (MITI) lends Japanese companies, at no interest, more than $500 million for research in high-technology areas. The operating style of Japanese companies has provided still another boost. Japan's booming electronics industry, the money machine whose revenues are used mainly to sustain still further growth, has come of age in a uniquely Japanese way: fierce competition between corporations has been preserved while at the same time the competitors have followed an overall cooperative plan. As the JTT article says:

> Back in the mid 1950s, Japan's industrial policy-makers targeted the 1980s as the decade when the Information Age would emerge in full bloom, with electronics as the basic technology that would power the economy into the 21st Century. In 1957 policy-makers passed into law . . . the Extraordinary Measures for the Promotion of the Electronics Industry.

In the following twenty-five years, Japan's electronics industry expanded by 150 times. It received a substantial boost two decades into that expansion as part of Japan's response to the oil crisis of 1973–1974.

In 1974, the Industrial Structures Council published a revised "Long Range Vision" of Japanese industry, designating the electronics industry as the key sector. As Michiyuki Uenohara of NEC said, electronics is "energy-saving, resource-saving, labor-saving and space-saving"—all virtues in an era of newly perceived limits.

In 1975, the Japanese electronics companies joined in a cooperative VLSI research-and-development project. In the following five years, the average productivity of the entire Japanese electronics industry increased by 2.57 times—an improvement by 21 percent per year sustained for the full five years and still rising rapidly.

FORTUNE 500 COMPANIES

The United States and Europe do not fully appreciate how well positioned Japan is, with its nationwide product-testing ground; its vertically integrated industries aiming for growth and value-added rather than quarterly profits; its cooperative rather than adversarial relationships between labor, management and government, to go on taking over more and more of the world's markets.

The degree of American insularity on these scores is even illustrated by *Fortune* magazine. For decades the magazine has made much of its *Fortune* 500 list of "industrial" corporations. But the list is largely a relic of a bygone era. More than half the names at the top of the list, 16 corporations out of 30, are those of oil companies. Pumping oil is certainly necessary, but it is not the industrial production of wealth. Industrial production must include the criteria that substantial value is added in the production process and that most of the materials used can be recycled, if not now then in a later, more sophisticated stage of industrial civilization. Industrial production means also that the products must be at a higher state of organization than the raw materials from which they come; in the language of physics, they must be at a lower entropy.

Steel production satisfies the criteria, while oil production does not. Rather than adding value, oil production simply depletes a nonrenewable mineral resource. Burning petroleum as a fuel increases entropy (disorganization) by taking complex, highly structured organic molecules concentrated in deposits underground and turning them, eventually, into gaseous carbon dioxide and various pollutants mixed throughout the Earth's atmosphere and therefore unrecoverable. Oil is not an industrial product because its price is determined not by productivity but by the decision of a few Arab sheikhs.

The *Fortune* 500 list also shows its age by listing only U.S. companies among the 500. Only in a later issue each year, and separately, does it list non-U.S. companies. When one eliminates the oil companies from the listing, and merges the American and foreign corporations, the result is far more meaningful in evaluating America's future in international competition. Only seven corporations in the world meeting my definition of "industrial" recorded sales of more than $20 billion in 1981. As of the early 1980s, American firms are still in the top four places. Listing annual sales for 1981 in billions of dollars, they are:

General Motors	$63
Ford	38
IBM	29
GE	27

Two of the remaining three among the seven biggest industrials are European:

Fiat (Italy)	25
Unilever (U.K.-Netherlands)	24
Du Pont (U.S.)	23

Du Pont's inclusion is due in part to its acquisition of the Conoco Oil Company in 1981. Considered on its chemical operations alone it would be in the next-smaller group, the eleven companies with sales from $15 to $20 billion per year. Of those companies, only one is American. I list them in decreasing order of size, and give nationalities by FR, France; D, West Germany; NL, Netherlands; CH, Switzerland; J, Japan; US, United States:

Renault	FR
Phillips	NL
Volkswagen	D
Siemens	D
ITT	US
Daimler-Benz	D
Peugeot	FR
Hoechst	D
Bayer	D
BASF	D
Thyssen	D

In that group one finds a few German companies whose names are not household words in the United States. Hoechst and BASF (the latter known to most Americans only by its tiny sideline of making recording tape) are chemicals manufacturers, each about three quarters the size of Du Pont and considerably larger than Dow. Thyssen makes steel and industrial products. The next group, of nine industrial corporations worldwide with 1982 sales in the $12–15-billion range, is dominated by the Japanese. Again listed in decreasing order of size, they are:

Nestlé	CH
Toyota	J
Nissan	J
United Technologies	US
Imperial Chemical Industries	UK
Nippon Steel	J
Western Electric	US
Hitachi	J
Matsushita	J

Although *Fortune* analyzes exhaustively the corporate performance of the 500 in terms of profits, it ignores the most important indicators for the future: whether a company is gaining or losing market share, whether its health derives from high value-added or is heavily dependent on raw-materials costs and whether its market is wide open or is near saturation.

Within the total list of the twenty-seven largest industrial corporations in the world, four (Unilever, Thyssen, Nestlé and Nippon Steel) deal with high-volume, relatively low-technology products. Four more (Du Pont, Bayer, BASF and ICI) are chemical companies, dealing largely with unchanging products and heavily dependent on petroleum as a feedstock for their factories. A ninth, Hoechst, is well positioned for the high-growth pharmaceutical industry but derives most of its revenues from commodity chemicals. Because those nine companies concentrate on high-volume, low-value-added products, spectacular new growth will be difficult for them.

Another third of the top twenty-seven are in automobile production. General Motors and Ford are the largest corporations in the world that fit my definition of "industrial," but they could be overtaken in a surprisingly short time. Many Japanese firms are growing at 15 percent per year, and at that rate Ford could slip to third place among the world's carmakers by 1989, and GM to second place three years later. Detroit's managers and workers are making unprecedented efforts to improve productivity and quality, but to succeed against the international competition they must bring off four revolutions simultaneously in a very short time. They must overcome a nationwide shortage of engineers that is not even generally recognized. They must penetrate new high-growth foreign markets where they have neither experience nor language preparation. They must dismantle or restructure the warring bureaucracies of management, government and unions. Finally, and toughest of all, they must resolve the difference in labor costs between Detroit ($20 per hour) and Kamakura ($12 per hour), either by producing 80 percent more per hour than the Japanese (which may be impossible) or by drastically reducing American wage rates (which so far no one has dared to suggest). General Motors decided in 1982 to begin importing subcompact cars from Isuzu Motors in Japan rather than try to manufacture such cars domestically. If there is to be high growth in the automobile industry, it is unlikely to be in the United States.

Of the top twenty-seven industrial firms, the remaining nine, split five-two-two among the United States, Europe and Japan, are all in high-technology areas: electronics, computers and flight hardware. All have the potential for high value-added and a comparative invulnerability to material and energy costs. They could grow almost without limit. The nine firms, in decreasing order of size, are IBM, GE, Phillips, Siemens, ITT, United Technologies, Western Electric, Hitachi and Matsushita.

PROTECTIONISM

The prospect for lower-technology, more conventional U.S. "smokestack-industry" firms is bleak. But the very size of the U.S. market, an economy at the $2.8-trillion level in 1982, suggests that Japanese competitors can only do better if we continue to give them free access to our market. The alternative is tariff-barrier protectionism. Yet even if we were to ignore international politics and our traditional free-trade policy, protectionism probably would not work. If we set up import barriers we will find our competitors retaliating in kind, and to survive as a major industrial power we must export more, not less.

There are two further pitfalls. First, so much of the world is becoming a market for sophisticated industrial products that our foreign competitors can do without us if they have to. That was brought home in the early 1980s with the A-300 wide-body airliner, produced by the European consortium Airbus Industrie. In its competition for sales in the United States against Boeing, Lockheed and Douglas it made almost no headway; its market penetration was limited to about a dozen planes sold to Eastern Airlines. But the market outside the United States had by then grown so large that Airbus Industrie was able to sell several hundred A-300s world-wide, achieving substantial penetration in a market formerly dominated by American aerospace firms.

Second, protectionism does more harm than good in the long run, even to the industries it tries to protect. Our threatened industries desperately need to catch up and surpass fast-moving foreign competitors in productivity, quality and the development of new products. Any industry we shelter behind the fragile wall of protectionism will stagnate. Japanese firms have been quick to exploit the possibilities for joint ventures. The Mitsubishi Electric Company, for example, has explored a potential joint venture with Western Electric to produce VLSI chips in the United States. Already, the Japanese electronics industry operates more than forty-five manufacturing or assembly facilities in other nations. In the United States alone, subsidiaries of Japanese firms account for 30 percent of all color television sets made, in addition to all the sets produced in and imported directly from Japan.

Since protectionism won't help us, it is tempting to probe for some other excuse not to compete with Japan on Japan's own terms. Competing head-on will require a painful transformation of our society.

There are changes going on in Japan that could relieve competition. The work force is aging, and some Japanese managers are pushing for a reduction of the retirement age for workers (but not for themselves) to 50 to save on seniority-based salaries. Small businesses that do subcontract work for big companies are finding it hard to keep their employees. Women are entering the labor market in greater numbers, and male-dominated Japanese companies aren't sure how to deal with them.

Yet even the Japanese may underestimate the ability of their industrial juggernaut to cope with change. One of Japan's most distinguished industrialists, Konosuke Matsushita, wrote a book called *Japan at the Brink* shortly after the 1973–1974 oil crisis. He reported that the government was so disorganized that members of the Diet were having fistfights instead of deliberations; that union demands for salary increases could kill Japan's export trade; that inflation was throwing Japanese society into chaos. But in fact Japan solved these problems far better than other nations during the following decade. It automated its factories, increased productivity, exported even more than before and as a result enjoyed a much lower inflation rate than its competitors. Japanese small businesses, once labor-intensive, are now automating rapidly to reduce their dependence on human workers and to give those who remain more interesting jobs. Many women go into computer programming, but as they reach the age of 35, past the peak age for programming performance, they generally retire to raise families and continue work at home on a cottage-industry basis at a lower wage.

INSULARITY AND CREATIVITY

The Japanese industrial machine has flaws, but we cannot count on their weakening Japan in the future. Flaw number one is insularity. For all their power on the international scene, the Japanese still do not think in global terms. The second flaw is an inability, so far, to nurture true creativity. Neither of those limitations is new. They fit the old-fashioned stereotype of the Japanese.

Japanese industrial managers are especially worried about Japan's long-term weakness in creativity. But as the Japanese are insular rather than global, the solutions they are working toward are isolationist rather than international. If those solutions are realized, Japan will become more nearly independent of America and Europe, and we will be left with even less to bargain with in our competition with the Japanese production machine.

I began my search for an understanding of Japanese creativity by interviewing several men who are actively responsible for the nurturance of creativity within their own companies. The first was Dr. Takashi Kitsuregawa of the Mitsubishi Electric Corporation.

The headquarters of the Mitsubishi corporations—Mitsubishi Electric, Mitsubishi Heavy Industries and the rest—form an enclave of tall ultramodern steel-and-glass buildings between the Victorian facade of Tokyo's main railway station and the walled, moated grounds of the Imperial Palace. While Kitsuregawa has served as Managing Director, his position at Mitsubishi is so high now that he is relieved of line responsibility. His personal signature on memoranda is the character "Kijirushi." It looks like a circled "K," and he uses it for speed. It can mean "eccentric." But his colleagues point out that it has another meaning: "genius." In Kitsuregawa's view, the United States and Europe are an idea factory, as powerful in creativity as is Japan in production. He said, "If we Japanese could not use foreign patents, we would be dead!" Unfortunately for the West, the basic research that nurtures ideas is a sink for money, not a cash generator like Japan's specialty, production. Obviously, the West will try to get the most it can for its ideas that flow to Japan.

Kitsuregawa realizes that Japan's creativity problem lies in Japanese society's nonacceptance of the creative individual. He pointed out that the Japanese historically were an agricultural people, using traditional, unchanging farming methods in a fixed land area without frontiers for emigration. In that environment it was difficult for any individual member of a social group to stand out above the others. Japanese had to be average to be socially acceptable; hence the often-quoted Japanese proverb "A nail that sticks out gets hammered down." Centuries ago Japan formed the habit of borrowing ideas from China and Korea, then Japanizing them. It happened with written language, religions, the arts and systems of government. During the modernization of Japan after the Meiji Restoration of 1868 the Japanese found it easier and cheaper to borrow than to create.

Kitsuregawa, however, did note times in Japan's history when creativity had flourished—the 8th-to-12th-century Heian period, and the 17th-century Genroku period, both times of extended peace when society felt so secure that it could tolerate the unusual. Creativity and innovation were also encouraged during the Samurai (Sengoku) period in the 16th century, but at that time wits had to be sharp to develop the strategics of survival. Creativity was again encouraged under the pressure of external threat centuries later. In 1932, Japan launched the Manchurian war to secure a source of iron and petroleum. Japan's government feared that with increasing prewar tensions the United States would cut off the flow of technological ideas, so it decided to encourage independent Japanese inventions. As a result, the Japan Society for the Promotion of

Science (Gakujitso-Shinkokai), government-sponsored, came into being in 1932.*

In Kitsuregawa's view, Japan already feels threatened, and is responding to the threat much as it did in 1932:

We Japanese are finding it difficult to buy patents abroad, because once we've bought the patents we can produce under license with better quality and lower cost than the patent owners. The West is hesitating to sell licenses, the prices for licenses are going up and often now the West insists on a barter of Japanese production methods in exchange for a new patent. That is why our Science and Technology Agency set up its Technology Creativity Development Program in 1981.

Kitsuregawa sees the Japanese solution to the problem in the development of homegrown Japanese creativity. "Although Japanese productivity and quality control are given a high mark today, it has taken the Japanese a long time to win worldwide praise for their products. By the same token, if they are allowed a little more time they will surely come to be recognized for their inventiveness and creativity as well."

I also visited Dr. Makoto Kikuchi, Director of the SONY Research Center, at the headquarters of the SONY Corporation to ask him about Japanese creativity. Its high-rise buildings are tucked away in the Kawasaki district of Tokyo, reached by a train journey and a walk through the narrow, twisting streets of a residential area. Dr. Kikuchi, slim and active, seems constantly in motion. His face mirrors great intelligence.

Dr. Kikuchi feels that creativity is widely misunderstood in Japan. True creativity, in his view, was displayed by Dr. Leo Esaki and his student Suzuki, who found the quantum-mechanical "tunnel effect" in semiconductors while working at SONY, and used it in designing the Esaki diode, an ultrahigh-speed electronic switch.

---

* The program paid off with the inventions of the magnetron, a device still used now for the generation of microwaves for radar, and of the Yagi-Uda antenna, used to generate a narrow, intense radio beam. The Yagi-Uda invention survives today as the familiar housetop television antenna, and the magnetron as the working element of the microwave oven. Ironically, in World War II the United States captured samples of both inventions. It then made far better use of them than Japan was able to, because in those days the American electronics industry was far superior to Japan's in production volume and quality.

And he agrees with Dr. Kitsuregawa that the magnetron and the Yagi-Uda antenna are the other outstanding examples of Japanese technical creativity in electronics.

But in Japan, he feels, the word "creativity" as used by Westerners is not really understood. When Japanese use it, it means what we would call "adaptive creativity," the ingenious development of a basic idea taken from abroad. He points out that the original meaning of the Japanese word *manabu* (to learn) is *manebu* (to imitate). When Japan's leaders in the time of the Meiji established the highly uniform, homogeneous structure of Japanese education, a structure that has endured to this day, it was to imitate the West and catch up to its level.

Kikuchi thinks that the Japanese will never—or at least, not for a full generation—be as individualistic as Westerners, and only rarely will create fundamental inventions. But Kikuchi feels that a Japanese student who works with a scientist in the United States for five years can become creative in the Western sense. American laboratories encourage creativity, and Kikuchi draws a lesson from his observations of them:

> Don't set up the lines of communication between researchers in a laboratory so well that everyone knows every wild idea that anyone else may have, because if you do, the good ideas will get squashed before they can gain strength.

He came back from a trip to the United States and tried to make that method work in his MITI laboratory in Japan. It failed, for the Japanese researchers all wanted to work together, and became very uncomfortable when separated.

Working cooperatively is a great asset once the basic direction of research has been set. Kikuchi recalls the late 1960s, when at conferences IBM scientists began referring to their "future system." They meant by it VLSI (Very Large Scale Integration). In Japan that sparked a rumor that the IBM researchers were developing a chip 12 inches across. False though the rumor was, it set fire to Japan's research toward VLSI, and the Japanese, working together, caught up very quickly.

Masaro Ibuka, now in his 70s, is tall, slim and vigorous, and retains the quick enthusiasm that he had nearly forty years ago when he founded the SONY Corporation. Ibuka is not yet satisfied that Japan, with half the population of the United States, already produces 50 percent more new engineers every year than does America. He thinks that the need for technically educated people is going to be still greater in the decades ahead, and he would like to introduce children to technology almost at birth.

Much of Ibuka's attention in recent years has been devoted to infant learning and infant education. With Glenn Doman of the Institutes for the Achievement of Human Potential, located in Philadelphia, Ibuka believes that schools in every nation start with children already past the prime learning years. Recently Ibuka has enlisted a number of families with infants to help provide him with data on how early infants can usefully be exposed to education. Six minutes of taped conversation in French is played to the infants once each day for several days. Then they hear no French until they are 2 years old, when they begin regular French-conversation lessons. Ibuka's data show that the children who were exposed to the sound of French when they were only 1 month old like it even better than those exposed to it at 3 months or 6 months. That kind of early education is, in his view, the right way not only to teach languages but to help introduce new, more accepting attitudes about creativity. He recognizes that education must begin very early in childhood if Japanese society is to overcome its traditional discouragement of the creative individual: "In Japan the personal family and the corporation are both family-like, and the acceptance of creativity must occur in both families if it is to be effective."

Matsushita's heavy investment in research suggests that Japan's creativity problems are more likely to be solved by corporate programs than by action in Japan's schools. The firm is Japan's largest single patent holder, with more than 100,000 patents in 65 countries. It invests more than $500 million per year in research and development, carried out by 10,000 scientists and engineers working in 24 different research establishments. Minoru Morita, Managing Director and Member of the Board at Matsushita Electric, told me:

Whatever anyone says, I personally don't believe that Japan will soon take over the lead in initial, basic technical development. The United States will continue to lead in that. . . . American society is very good at stimulating individual creativity—finding the seed and nurturing it. Japan must try to create the right atmosphere for creativity.

He feels that America's problems in competition with Japan come from failings in U.S. society that will be as difficult to overcome as Japan's problems with creativity: "American companies are not good at production, because Americans are not good at working as a group."

Hiroshi Watanabe, Executive Managing Director at Hitachi, Ltd., is as critical of the Japanese university system as are other senior executives. He acknowledges that Japanese industries lead the world in productivity and reliability, but says:

> If we go one step inside, Japanese ability still seems very shallow —as in the design of and basic research on circuits and subsystems, for example. Japanese universities today keep themselves in many respects within the traditions of the Meiji era. There is not a single university in Japan that has facilities for college and graduate students to make LSIs.

And looking toward still newer products, he said:

> It takes twenty to thirty years for a science to develop into a big industry. So when we look ahead for an age beyond electronics, the first important thing is to identify sciences that will be supporting the postelectronics era. Yet our universities are quite indifferent about providing help to basic research.

The Hitachi Technical Institute, a large complex of buildings hidden in a maze of narrow residential streets, makes the best of the industrial and academic worlds by hiring university professors to do its most advanced teaching. The Institute began seventy years ago, training workers fresh out of junior high schools. In 1960 it added a college-level program for high school graduates. And ten years later it began postgraduate education for engineers and scientists, staffed by the best teachers it could find in Japan's universities. Clearly, the Hitachi Institute is planning for a time when Japan will no longer have to barter production expertise for inventions.

Occasionally in Japan one finds an individual whose creative ability seems cast more in the American or European than in the Japanese mold. There will be many more if infant-learning programs like Ibuka's, corporate programs like Hitachi's and government efforts like the Technology Creativity Program succeed. One such man is Dr. Kesaharu Kasuga, President of the Nitto Boseki Company, Ltd. Nitto Boseki's developments in fiber-glass production are especially innovative. Dr. Kasuga, winner of several awards from industry groups, is proud that "Nittobo" worked out its methods independently; all similar Japanese firms still manufacture under license from the United States or Europe. To a considerable degree, under his direction Nittobo has already become an idea factory. It licenses its methods to other countries—the United States, West Germany, Brazil, India and others—for a down payment plus a 3-percent royalty against sales. The royalties earn for Nittobo 28 percent of its total profits, a higher percentage from royalties than is earned by any other Japanese firm.

Kasuga gave figures for Japanese industry as a whole that showed how quickly Japan is developing an independent creative capability. While the Technology Creativity Program formally began only in 1981, the government's efforts to improve creativity started ten years earlier. At that time Japan was paying 100 times as much for licenses imported from abroad as it was earning on the sale of export licenses. But by 1981, import and export license payments were almost in balance.

Dr. Kasuga is proof that Japan can eventually provide the necessary creative base for industry without help if it has to, making Japan an even more formidable opponent. That being so, it is some comfort to know that at the very highest levels in Japan there are men who feel a worldwide sense of responsibility.

Konosuke Matsushita, like his younger counterpart Masaro Ibuka, thinks in global terms and has devoted his later years to education. He has said that

> I decided that the task of the industrialist is to make his products widely available at the lowest possible cost to bring better living to the people of the world. Profit should not be the goal of a business enterprise. The primary goal of business should be to contribute to society in return for having had the use of society's resources.

These views of Japanese and American industry by our strongest competitors are enough to shock us out of any complacency we may have had left. The sad performance of American industry in the

1970s, in competition with Japan, stems from failures of our own that are clear enough to the Japanese. Our educational system starts too late and badly neglects the "hard" subjects like mathematics, engineering, science and non-European languages—just those subjects which young graduates need most if they are to contribute to high-technology industries competing in world markets. Management fashions—sacrificing long-term growth for quick profits, glorifying the merger and the takeover as if our industry were a nationwide board game—ignored productivity, the source of real wealth. Adversary relationships, government/industry and management/labor, survived long past the time when they were justified, and left us unable to compete with the united, cooperative industrial machine of Japan. And our own American brand of insularity, preserved in such anachronisms as the *Fortune* 500 listing, kept us from recognizing our problems until far too late.

As we have gradually come to appreciate these realities in recent years, we have still taken comfort in the assurance that for all their superiority in production engineering, the Japanese would lag in basic invention. But as we have seen, they are working on that also. Nurturing true creativity within the Japanese society will require a major transformation in their culture—but it is no harder than the transformation we must carry out in our own. Whether or not the Japanese succeed in their transformation, we cannot afford to fail in ours.

Now, in Part II, we will investigate the market opportunities of this next decade. As we review them, we should prepare ourselves to seize those opportunities—lest they too be lost.

Two

SIX MAJOR OPPORTUNITIES

OVERVIEW

In the traditional "smokestack industries," making automobiles, steel, commodity chemicals and similar products, the international battles for market share began long ago. The United States has already lost some of those battles, and it may lose more. None of those industries offers opportunities for the United States to recoup its economic fortunes in competition with Japan and Western Europe.

But newer, higher-level technologies have opened market opportunities that the United States—or any nation—could still exploit for major economic growth. For true success, however, the three criteria for growth that I set out in the introduction to this book must be met: better service, better energy efficiency and less damage to the environment. Even that is not enough. Americans need to know that a new industry can grow, and that the United States can share in its economic growth.

An industry, existing or potential, must satisfy three conditions if it will soon help the United States. The science and engineering for it must be well established, so that no long, painful, uncertain period of research is needed before it can generate revenues. Yet the technology must be new enough, and developing rapidly enough, that competition remains open. Finally, the potential market must be large enough to make a real difference to the American GNP and balance of trade.

Six candidates appear to satisfy all those conditions. We will determine which ones offer real promise for the next decade and which, attractive as they may be, cannot spawn major industries until late in the century. The opportunities progress from the most familiar and mature to the least-known. The microengineering industry is already generating more than $100 billion per year, doubling every four years. Self-replicating robotics is a younger industry, but growing even faster. Genetic reconstruction is well established, but its potential is still controversial. Three less familiar technology areas afford opportunities to open major new industries whose benefits would transform our lives. Those areas are magnetic flight, personal aircraft, and large-scale construction in space. In magnetic flight the United States seems hopelessly outclassed—but it still has a chance to win, if it leapfrogs technologies that Japan and West Germany have developed. The opportunities for major new growth markets in personal aircraft and space construction remain untapped by any nation.

It is foolish to rush blindly into any new-technology area without knowing its potential pitfalls. I will begin with three cautionary tales, each a history of how a nation or an industry lost out disastrously in what looked, at first, like a great opportunity.

In the first case, the basic technological idea was sound, but the nation where it was born failed to exploit it vigorously and effectively.

On the fifteenth of May of 1941, Frank Whittle, a junior RAF officer not yet 34 years old, stood on an airfield in Lincolnshire as the first Gloster E-28 prototype jet aircraft lifted off the runway and screamed its way to cloudbase. At the moment of lift-off, a friend slapped him on the back and yelled, "Frank, she flies!" Whittle, who had spent fifteen years designing the jet engine and persuading a largely hostile Air Ministry to take it seriously, snapped back, "That's what it was bloody well designed to do, wasn't it?"

Unknown to Whittle, the German Heinkel-178 jet research aircraft had already flown in August of 1939. In the early 1930s, Whittle had patented his work but asked that the patents be kept secret. The Air Ministry had refused, the patents had been published—and Germany had raced to develop jet engines. As a result, the Messerschmitt-262 twin-jet fighter-bomber later became the only effective jet aircraft to see service in World War II. When the war ended, occupied Germany was in no position to compete, but in the United States both General Electric and Pratt & Whitney, having been given Whittle's drawings and the engines themselves, developed the jet to far higher levels of thrust and efficiency than

Britain's weaker industrial establishment could manage. Then the worst blow to Britain's jet-engine prospects occurred: the loss of Whittle himself. His own small firm, Power Jets, was nationalized and prohibited from building jet engines. Whittle, who had already invented the bypass turbofan engine that was to be reinvented years later, resigned in disgust and ill health.

In the following years Britain's jet-engine industry was eclipsed by American firms, which were able to develop engines under Air Force contracts for the military precursors of later civil transports. An attempt years later by Rolls-Royce to reenter competition with a superadvanced engine for the Lockheed L-1011 Tri-Star resulted in bankruptcy for Rolls-Royce and in a loss of market share by Lockheed from which the L-1011 never recovered.

The second cautionary tale is of a different kind. Makoto Kikuchi of SONY tells it as an example of "the risks that the pioneers take. They pay a heavy tax for being first." In the 1960s, companies like RCA and GE in the United States thought they could use an obscure principle of physics called the Peltier Effect to open up a new market. The physics was sound: in a thermocouple, two wires made of different metals are welded together at two points, forming junctions. When one junction is put in a warm place and the other in a cold, heat energy is converted to electric energy and a current flows. Thermocouples are the basis for most industrial temperature-measuring devices. A thermocouple can also be run in reverse: if one forces current through a thermocouple when its junctions are at the same temperature, one of the junctions will grow hotter while the other grows colder; that is the Peltier Effect.

The American engineers had their sights on a refrigerator that would require no motor or compressor. In attempting to develop such a machine, based on the Peltier Effect, they spent more than $50 million. It was a waste of money. The materials, like copper, required for a Peltier refrigerator became more and more expensive as the years passed. Meanwhile, conventional motor/compressor refrigerators became less and less expensive as automation was introduced into the factories where they were built. Ultimately, the project was abandoned.

The third cautionary tale is of the same kind, but it is a story not yet finished. Like the second, it illustrates the penalties for violating the three basic criteria.

Since 1967, General Motors has been working toward a practical electric car. In 1979, GM announced a "breakthrough" with its development of a somewhat higher-performance battery using zinc and nickel oxide. But in 1982, Roger B. Smith, Chairman of GM, admitted the electric car was "on the back burner." It lost priority because gasoline was plentiful at that time, and because of dramatic advances in gas mileage achieved by ordinary internal-combustion engines when their designs were improved and their carburetors and spark timing were controlled by microchip computers. GM hasn't totally shelved the electric car, but Hideo Sugiura of Honda Motors said:

> There won't be any replacement of the internal-combustion engine in this century to more than a five-to-ten-percent level of market penetration. There will be a lot more improvement of engine performance, especially efficiency, through microprocessor control and through new combinations of old ideas like the turbodiesel. But electric cars won't equal the performance of internal combustion for a long time, and we won't be able to sell electric cars until their performance *is* competitive.

In those two examples of "technological revolutions that never happened," the technologies failed the first of the basic three criteria: they did not provide better service. In the chapters that follow, the three criteria will be applied to each of the opportunities that lie ahead, and we will identify those nations which have the best chance of dominating each new industry.

CHAPTER 2

Microengineering

Of the technological changes that have altered our lives in recent years, no other is so pervasive, dramatic and powerful as the information revolution. Expressions like "Information Age" and "Knowledge Era" are commonplace. The hardware that now processes digital information would have been unthinkable less than two decades ago: powerful computers available for the home at the price of a hi-fi set; digital calculators no larger than a business card; clocks and watches with no moving parts other than a vibrating quartz crystal. The breathtaking progress in the handling of information with ever-greater speed and in vastly greater amounts has come about because the basic unit of information, one bit, has no intrinsic weight, size or energy.*

There is almost no limit to how much information can be stored in a given volume, and almost no limit to how many bits per second can be transferred from one place to another. That is why the race to develop information machines of greater and greater capability is a race to miniaturize, and why its vehicle is microengineering. The development of microchip electronics is an important part of that race. Only one hard limit—the speed of light—rules in the world of information. In this next decade, for the first time, our most exotic information machines will approach that limit.

* A bit, the abbreviation for "binary digit," is a numeral in the binary (base 2) system. It can be a 1 or a 0. The binary number corresponding to the decimal 19, for example, is 10011. In electronics, the binary 1 and 0 correspond to two different voltages.

THE ELECTRONIC MICROCHIP

The transistor circuits on semiconductor microchips are of two different kinds: memories to hold bits of information, and "gates" which process that information, performing the operations of arithmetic and of logic. The memory in the typical personal computer of the early 1980s, the Apple II Plus, was made up of two dozen "16k" chips, each able to store 16,384 bits of information. (The number is 2 multiplied by itself 14 times, because computer arithmetic is based on the number 2 rather than on 10.) In chip technology, an advance by one "generation" means a reduction of the scale of all transistors and interconnecting printed-circuit conductors to half the length and half the width of the previous generation, quadrupling the number of circuits per chip. In 1980, Motorola became the first company to advance memory technology by one generation beyond the 16k, to VLSI—a 64k chip with 65,536, or 2 to the 16th power, bits of information.

Motorola believed that "Memory is where the money is," and the Japanese knew it too. NEC, Hitachi, Toshiba and several smaller firms jumped into the market, and quickly built production facilities that were more highly automated than the Americans', particularly for the crucial bonding process (the welding of metal wires onto a chip to connect it to external circuits). Only the largest-volume, most highly automated producers are able to compete in the field of memory-chip production. In just the two years 1981–1982, competition pushed down prices from $20 to $5 for a 64k chip. Memory manufacturing is definitely not an easy, get-rich-quick business. As Hiroshi Asano of Hitachi said, "Initially we thought this product would be a supergolden egg, but it's turned out to be just a plain golden egg." Of the U.S. firms, only Motorola and Texas Instruments were left in the 64k market by 1982, with 70 percent of the 64k production Japanese.

The American companies tried to leapfrog by advancing another generation, to chips with 256k bits, but Hitachi and NEC kept pace, beginning quantity production of 256k chips in late 1982. Dr. Uenohara of NEC expects the price of a 256k memory chip to drop as low as that of a 64k fairly quickly. But lowering prices raises the total market so dramatically that the market for all types of chips, already over $10 billion in 1980, is doubling every two years and is expected to continue doing so through the decade. By the mid-1980s, nearly all personal computers and many other consumer products will be using 256k memories. To get an appreciation of what that will mean, it would take only fifteen VLSI 256k chips, each smaller than a fingernail, to store all the words in a four-hundred-page book.

The reason that manufacturers can produce low-power, hand-held computers which run on small batteries, while the ordinary personal computer must be plugged into an electric socket, is that there are two quite different types of transistor, both in wide use. One goes by the name N-MOS—a metal oxide–silicon chip in which negative charge carries electric current through each transistor. N-MOS transistors are fast, but they require a good deal of power. The C-MOS is better suited to small battery-powered calculators and computers. The "C" stands for "complementary," and it means a combination of an n-type (negative-charge carrier) and a p-type (positive-charge carrier) transistor in a pair which functions like a canal lock. Current flows, and power is drawn, only during the instant that the output of the pair switches between its high- and low-voltage states, corresponding to the 1 and 0 of binary arithmetic.

C-MOS is particularly well suited to memory chips, which spend most of their time quiescent rather than switching. By 1981, Hitachi and Matsushita had found a way to combine N-MOS and C-MOS on a single chip, with low-power C-MOS memory units near the center and fast N-MOS switching gates at the edges. To save power, these chips switched off the N-MOS gates when those gates were not in use. Motorola bought Hitachi's new "merged" N-MOS/C-MOS technology as the basis for a more advanced generation of calculators. Dr. Uenohara of NEC expects 256k memory chips to be made by the new N-MOS/C-MOS technique. When they become available, the memory capacity that once required a plug-in, tabletop computer can be provided in a new generation of highly capable hand-held computers.

MICROPROCESSORS
Impressive as are the advances in memory chips, the stars of the chip galaxy are the microprocessors. These computers-on-a-chip combine a great many logic gates to form the equivalent of a large computer's central processor unit (CPU), the component that does the thinking, as opposed to the remembering, in a computer. The capability of a microprocessor can be measured roughly by its

"word length," the number of 1-or-0 bits that it can handle simultaneously. The first microprocessor, Intel's 4004, introduced in 1971, could deal with a 4-bit "word." The spark for the explosion of personal computers like the Apple II and the TRS-80 was a new generation of 8-bit microprocessors that followed three years later. With just two operations those microprocessors could handle a full 16 bits, enough to specify 64k locations in the computer's memory. A microprocessor design tends to have a long useful life, because the computer-on-a-chip gets incorporated into a great many larger, more complicated products, from personal computers to hi-fi stereo receivers. And with increasing automation, the price of microprocessors has dropped spectacularly. Intel's 8-bit 8080 cost $360 when it was first announced in 1974. By 1976 it was down to $32, then to $6 two years later, and by 1982 its price was under $3. In 1982, world production of microprocessors was 200 million, equaling the total made in all prior years combined. By 1990, there will be from 3 to 4 billion microprocessors produced each year, and a home might have twenty or thirty, running everything from energy-efficient refrigerators to home communications and alarm systems.

In 1981, the microprocessor art advanced by another generation, to 16-bit chips like Motorola's 68000 and Panasonic's MN-1610. Those devices were at the heart of second-generation personal computers like the IBM "PC," which used an Intel 8086. By that time, Intel and its competitors were making specialized automotive computers that from a single chip controlled interior climate and cruising speed, detected knocking, read out the time, kept brakes from locking and combination door locks from being broken into and maintained tuning of both the car engine and the stereo radio.

Toshio Takai, Executive Vice-President for the Electronics Industries Association of Japan, expects that personal computers "will replace consumer electronics [stereo, TV, VCRs and so on] as the driving force behind the semiconductor industry." Microprocessors constitute about 6 percent of the semiconductor industry, and through the early 1980s they remained largely an American preserve, though some Japanese manufacturers are confident that they will be taking a much larger share of the U.S. microprocessor market in the near future. Indeed, in at least one area the Japanese have already done so. Just as in the case of memory chips, microprocessors can be made with low-power C-MOS technology. The Sharp Corporation in Japan, under the leadership of Atsushi Asada, made a large investment in assembly lines for C-MOS microprocessors on the bet that it would open up a market for truly portable personal computers. No U.S. manufacturer was in a position to make a similar investment, so the Sharp Corporation, now experienced enough to be able to produce at low cost, commands a lead.

Numbers like "4-bit" or "16-bit" need connections to real-life tasks before one can be comfortable with them. Roughly, a 4-bit microprocessor is adequate for a pocket calculator; an 8-bit microprocessor will do an excellent job in a word-processing program (if the program is well written and efficient); and it takes a 16-bit microprocessor to do a good job on high-resolution color graphics. When we think of color graphics we tend to think of computer games, but 16-bit microprocessors are now being used for more serious color-graphics applications: in medicine, the detailed imaging of internal organs from ultrasound or X-ray data; in industry, the visualization of process control and assembly-line operations.

During the 1970s, the typical industrial or scientific minicomputer was a 16-bit machine. But there are important applications that demand still more computing power, particularly computer-aided design and manufacturing (CAD/CAM, as designers call it now). In CAD/CAM, the designer needs to be able to compose complicated three-dimensional images in the computer's memory, and have them displayed on his two-dimensional screen from any viewpoint and with any magnification he chooses. To do CAD/CAM properly, one really needs a 32-bit machine, and up to the late 1970s that meant a large "mainframe" computer. Then the Digital Equipment Corporation (DEC) brought out its VAX-780 series of "supermini" 32-bit computers, and they were extremely successful.

After several years of work that cost from $30 to $40 million, Intel in 1982 began shipping its Model 432 "micromainframe," a 32-bit microprocessor on just three chips that together could fit on one fingertip. Intel's commitment to the micromainframe project was made in 1975 by its Chairman of the Board, Gordon Moore. At that time the semiconductor industry was in a recession, and while other companies were laying off computer scientists, Intel was hiring the brightest it could find. To run the new program Intel brought in William Lattin, an experienced engineering project manager, from Motorola. Justin Rattner, then only 27, became the design leader of the micromainframe project. Partly to maintain secrecy, the project was relocated from California's Silicon Valley to the suburbs of Portland, Oregon, in 1977.

With support from Lattin, Rattner incorporated several state-of-the-art features into the 432 rather than simply shrinking an existing design. Conventional computers handled data in the form of individual numbers. Instead, the 432's "architecture" (logical design) allowed it to handle blocks of data as single entities or objects. That "object-oriented" architecture, which programmers could make use of through an advanced language called "ADA," sealed the computer's data against unauthorized access or manipulation much more effectively than older, number-oriented architectures.

Rattner also incorporated into the 432's design the ability to perform several computing tasks simultaneously, and made that multiprocessing capability "transparent," which meant that several 432s could be hooked together for greater power without the need for reprogramming. Some of the 432's operating programs were permanently stored in the chips themselves, and because of transparency, a system made up of two or more 432s could continue to operate, though at reduced speed, if one of the micromainframes failed entirely.

The 432 is finding its main uses in factory systems controlling as many as fifty robots all from a single station, in telephone switchboards and in new families of supermini computers that do for about a quarter of the price tasks that previously required a quarter-million-dollar VAX-780. Digital Equipment, the manufacturer of the VAX-780, is rushing to produce a micromainframe version of that machine, and IBM is sure to be working on something similar. Intel regards the 432 as the grandfather of a series whose useful life will extend into the 1990s. The original version had about 250,000 transistors on its three chips, and the later generations will have 4 to 7 times as many. A 432 sells for only a few hundred dollars, but microprocessors are the fastest-growing segment of the semiconductor market, and they have a strong "drag effect," pulling in large orders for auxiliary chips. Intel expects micromainframe-related sales to approach $1 billion by 1990.

Microprocessors are far more complex in their design and more related to the subtleties of system architecture than are memory chips. The micromainframes are another giant step in complexity beyond previous microprocessors. For those reasons, it is not very easy for Japanese companies to leapfrog ahead of them with still more advanced designs. Intel has been the pioneer and world leader in microprocessors, and its designs usually become industry standards. Through the 1980s it seems likely that Japanese companies will continue to use the "scavenger strategy" of copying U.S. micromainframe designs, probably with enhancements to make them more usable in systems that communicate in ideographic characters. Moore of Intel is hopeful that his firm will be able to stay ahead: "The technology of copying doesn't seem to be improving as fast as the new products. People even had trouble copying our 8086 microprocessor."

MILITARY AND DEFENSE USE

Many Japanese feel that the United States lost the race for the consumer-electronics market in part by being preoccupied with military hardware. Department of Defense officials became active in the "chip wars" again in the 1980s because of their constant worry over Soviet superiority in armed-service manpower, tanks and missiles. They wanted to offset that superiority by making U.S. weapons "smarter" than their Soviet counterparts. Equaling the technical level of U.S. and Japanese commercial equipment was not enough, because what the Japanese could make, the Russians could buy through intermediaries.

The DOD solution was to sponsor, late in the 1970s, the VHSIC (Very High Speed Integrated Circuits) program. Where the Intel micromainframe put a quarter million transistors on three chips, the VHSIC program aimed to put a million transistors on one chip, advancing the electronic art five to ten years beyond what the DOD thought the Soviets would be capable of. The result would be a single family of circuits that all the armed services could use. Those chips would have 100 times the speed of existing circuits, and so much redundancy, versatility and self-diagnostic capability that they would continue to function accurately even if partly damaged. Norman R. Augustine, Chairman of the Defense Science Board, predicts that VHSIC microcomputers will give missiles computing power "in the same league as that of the human brain." Such microcomputers could lead to "brilliant" rather than merely "smart" fire-and-forget missiles, defensive radars for new-generation fighter planes and new fire-control systems for tanks. To achieve the extreme speed and circuit density required in the VHSIC program, the companies participating in it have to develop circuits whose conductors are only 1.5 microns (thousandths of a millimeter) wide. Such conductors, if combined, would form a cable only as thick as a human hair, but with a thousand independent wires in it. Honeywell, Hughes, IBM, Texas Instruments, TRW and Westinghouse all have major roles in the VHSIC program, which is scheduled to complete a $200-million development phase in 1984. If on time, it will begin to provide new-generation computers for weapons systems by 1986.

To translate the new chip technology into military hardware, the Defense Advanced Research Projects Agency (DARPA) set up a new facility that links designers at CAD consoles in several dozen companies and universities by "Arpanet," a satellite communication system that is encrypted for secrecy. The network permits rapid sharing of software, transmission of chip circuit designs and consolidation of new designs at one laboratory for implementation at a "silicon foundry." In some cases, new concepts make their way through the new facility from design to custom fabrication in as little as two weeks.

DARPA has a leading role in another new-technology development. Silicon replaced germanium in the 1960s as the preferred element for semiconductor chips. Now a still newer material, gallium arsenide, is coming into limited use for special purposes. Gallium arsenide, GaAs, is a metallic compound of arsenic and the rare silver-white metal gallium.* No one could make it pure enough and uniform enough in quality for semiconductor use until a small company, Metals Research, Ltd., developed a new machine called a "Melbourn crystal puller." Even afterward, GaAs remained 15 times as expensive as silicon, which limited its commercial use. But GaAs has advantages over silicon that have drawn DARPA support for its development. The speed of electron motion in GaAs can be six times as fast as in silicon. In practical operation, that translates into a GaAs computer which can run twice as fast as its silicon counterpart; and because of the extra speed of the electrons, a GaAs device can operate at much higher frequency than a silicon one.

For the first time there is the possibility of replacing tubes in high-frequency radars with more reliable solid-state circuits. GaAs can run at a higher temperature than silicon, and—important for missile-guidance applications—it can tolerate more intense radiation without being damaged. GaAs-chip technology is still in its

* Semiconductor elements are made up of atoms whose outer electronic shells are neither filled, like those of the noble elements argon and krypton, nor almost empty, like those of the alkali metals sodium and potassium. Germanium behaves chemically like the more familiar elements silicon and carbon.

infancy; not until 1980 could even 1,000 circuits be put on a GaAs chip. But the new material has potential applications to large civilian markets: receivers for direct satellite-to-home television broadcasts, and amplifiers for fiber-optics communication systems. In the United States, progress in GaAs-chip technology is paced mainly by DARPA's funding, but already there is concern that Japanese companies, which are putting more investment into nonmilitary applications of GaAs, may be ahead. At the Fujitsu plant where Japan's largest computers are developed, much of the conversation focuses on using GaAs chips for the highest-speed sections of future computers.

The Defense Science Research Board called the VHSIC program the single most important technology development now being carried out by the U.S. military. But observers question its benefit to the U.S. electronics industry in our peaceful competition with Japan. Some specialists have charged that the VHSIC program has siphoned off high-quality designers who would otherwise have been available for that competition. And so far no one has thought of an effective way that the VHSIC technology could be transferred to nonmilitary U.S. industries without its being leaked to the U.S.S.R.—giving up just the advantage for which the program was targeted. As for the headway it will give U.S. manufacturers over the Japanese, a lead of as little as a few months in the fast-paced semiconductor business has often meant the difference between dominating a new market and losing it completely.

Perhaps the most significant beneficial effect of the VHSIC program is the example of cooperation it has provided for U.S. industry. As Gene Strull of the Westinghouse Advanced Technology Division says, "The VHSIC program is the first real America, Inc., with firms working together for the betterment of the United States." Considering what the American electronics industry is up against in the way of commercial competition, it probably needs a nonmilitary version of the VHSIC organization as a regular working tool to hold market share in consumer products.

CONVENTIONAL TECHNOLOGIES

Sho Masujima of TDK Electronics predicts that by the mid-1980s the standard-size audiocassette will pack four times as much recording time as it does now, because recorders will begin using a crosswise raster scan instead of the old-fashioned straight along-the-tape recording. And the floppy disk is being made in smaller and smaller sizes, with fierce competition over standards for each new size. IBM introduced the first floppy disk, of 8-inch diameter, in 1970, and its format became an industry standard. In 1974, Shugart Associates brought out the 5.25-inch drive, and its many imitators adopted the Shugart standard. In 1982, SONY introduced a new, smaller 3.5-inch model. Matsushita and Hitachi countered with a 3.0-inch version. To confuse the issue further, several U.S. firms led by Shugart banded together to establish still another standard. But since none of them had a marketable product, they were accused by the Japanese of trying to "form an OPEC without any oil," with their suspected purpose being simply to slow down the Japanese.

A very large market can be opened, without any new technology, by the introduction of digital broadcasting, receiving, recording and playback systems, for both television and audio. People who use personal computers are getting used to the 100-percent fidelity of digital systems, and as a result are becoming dissatisfied with the generally poor technical quality of television, which depends on analog signals.

The transmission of analog signals is very much like telling a story to a friend, then having the friend tell it to another friend, and so on in a chain. In each link of transmission, and with each stage of copying, some of the quality is lost. Digital systems encode the original analog information into a sequence of 1s and 0s (binary bits), and at each link in transmission or copying, the receiver needs only to recognize the difference between a 1 and a 0, then re-forms a perfect binary bit before relaying it on. In 1981, Intermetall, a West German subsidiary of ITT, demonstrated the first digital color television system at a trade fair in West Berlin. During the late 1980s there may be a new round of the consumer-electronics battle as hundreds of billions of dollars' worth of analog TV and audio systems in homes are replaced by digital versions. The American and European electronics companies have a chance to upset their Japanese rivals in that new battle; but they will have to organize it far better than in the first, "analog round," or they will lose a market even more lucrative than that of the VCR.

"OFFICE OF THE FUTURE"

The biggest potential market for the 1980s is the "office of the future." Even in personal computers, 60 percent of all purchases are for business use. The information-processing market in the

United States alone in 1982 was $48 billion, and the office, large or small, remains a tempting but intractable target for automation. About 60 percent of the $1.3 trillion spent on wages, salaries and employee benefits in the United States in 1980 went to office workers. It is much harder to measure productivity in office work than in a factory, but the continual increase of "overhead" charges in every organization suggests that office productivity is improving only slowly if at all. The few really successful examples, like airline reservation systems, all depend on very large central computers linked to hundreds or thousands of work-station terminals. There the industry giant is IBM, and when Japanese firms were given their assignments by MITI for the assault on the U.S. computer industry, Fujitsu was the company selected to compete head-on with IBM's biggest machines.

Fujitsu's latest and biggest computer, the FACOM M-382, is tough competition for IBM's 3081K. The M-382 is twice as fast as the 3081K and significantly smaller, occupying only about 60 percent as much floor space. It is a useful advantage in marketing that the Fujitsu machine can simply be air-cooled, while the IBM needs water or oil. Fujitsu is marketing the M-382 aggressively in the United States, Western Europe and Australia.

The strategy not only of Fujitsu but of all Japanese firms competing up and down the range of IBM's product line has been to pick a target and try to build something with slightly higher performance, fully compatible with IBM software and peripheral hardware ("plug-compatible"). The Japanese are willing to sell their new product at a substantial loss, sometimes as much as an 80-percent discount, in order to get a customer to switch away from IBM. The American firm is well aware of the strategy, and in recent years has counterattacked by developing each new product in greater secrecy, thoroughly automating its production line before the first sale and then putting the new product on the market at a price so low that Japanese competitors could well go broke trying to match it. IBM made that play with its 4300 series of business computers, and complemented it by deeply cutting, in some cases by as much as 50 percent, the prices on CPUs that were bumped to a less than top-of-the-line position by newer models.

SCIENTIFIC COMPUTERS
Because IBM concentrated so effectively on the business-computer market, newer American firms looked for opportunities in the field

of scientific computers. At first it seemed that the specialized scientific machines they developed would have relatively little commercial impact; but during the 1980s those machines have turned out to be extremely valuable in the largest aircraft and automobile companies. They combine hardware and program capabilities that have evolved from forty years of computer development.

In the 1940s, the Harvard Mark I computer was built with sponsorship by the U.S. Navy. Grace Hopper, who later became the head of the Navy's Programming Languages Department in the Pentagon, worked on it and developed the first "compiler," a program that accepted English-like commands input at a keyboard and turned them into machine language that the computer could understand. From that came COBOL—Common Business Oriented Language. Over the following decades, the trend in communication with computers has been toward compilers of higher and higher level—that is, more English-like and easier for nonspecialists to work with. A single command made in the language of a modern high-level compiler like PASCAL may generate several hundred machine-language instructions.

AERODYNAMICS

As more and more sophisticated compilers were written, allowing even nonspecialists to continue directing machines thousands of times faster and more complex than the Mark I, tougher problems were tackled, always up to the limit set by the speed of computers. In the 1960s, the Control Data Corporation (CDC) developed its 6000 and 7000 series computers, in which state-of-the-art transistor circuits were directed by formula-translation (FORTRAN) compilers to do the gigantic computations necessary for the analysis of elementary-particle-physics data. But until the 1970s, no machine was fast enough or could remember enough numbers to tackle the big problems of aerodynamics: solving for the compressed flow of air over the complicated shapes of aircraft moving near or above the speed of sound.

Then came a new concept in computer architecture. It has several different names: array processing, vector processing, pipelining. The meaning is the same: The hardware of the central processor itself is designed around the most time-consuming tasks that the computer will be called upon to do. Specialized hardware is built in to do very quickly the computations that are both lengthy and most often needed, and the rest of the CPU is set up to anticipate when those computations will be needed. The supporting circuits feed input data to the specialized hardware as soon as it is ready for them, and then buffer it by carrying out other tasks while waiting for the number cruncher to finish its job.

In the 1970s, the CDC STAR and the CRAY-1 computers built with the pipelining concept achieved speeds of up to 90 million calculations per second. They were used for aerodynamic modeling that would have been impossible earlier: calculation of the lift and drag of the piggyback combination of a Boeing-747 and the Space Shuttle Orbiter; design of the "supercritical" wings of the Boeing 757, the Airbus 310 and a new generation of fast, fuel-efficient business aircraft. Fujitsu's M-382 uses three separate levels of buffering in its pipelining, so that the core of its central processor runs at an almost steady rate, protected from the varying demands of the computation going on several levels above it.

There remain problems too big even for computers in that class. To wring the last ounce of efficiency out of aircraft, with potential savings of many billions of dollars per year in fuel costs, computers must be able to trace the growth of turbulence as the boundary layer of air flowing over a wing separates from smooth laminar flow; they must be able to trace the interference of shock waves with the aircraft structure; they must be able to predict the combined oscillation of compressible air with nonrigid, elastic wings and control surfaces. To tackle such problems, the Numerical Aerodynamic Simulator (NAS) now being designed for NASA will use "conventional" silicon chips, but it will combine them in several CPUs, each with parallel processors using pipelining. It will perform a billion operations per second, ten times more than the CRAY-1, and it will have rapid access to a memory of 240 million multibit words. Japan's National Aerospace Laboratory is keeping pace with a comparable machine, to be operational by 1990 at a cost of $150 million. The fuel bill for just five wide-body airliners equals that price in the course of a year.

MEETING THE SPEED OF LIGHT
To go still faster—and there are plenty of design problems that need even greater speed—computers will have to meet the limitation of the speed of light. A NAS-type computer can do a calculation in the time it takes a data signal moving at the speed of light to

travel just one foot. So to calculate even faster, a computer, including both its CPU and its rapid-access memory, must be made smaller than a briefcase. To obtain the necessary speed, silicon chips must be abandoned for more exotic solid-state devices; GaAs is a first candidate, but designers think they'll have to go to something still farther out: Josephson-junction circuits. Josephson junctions depend on superconductivity and must therefore operate at a temperature of almost absolute zero.

The Josephson junction in its most common form is a pair of superconductors separated by a very thin insulator—typically, 1 millionth of a millimeter. Philip Anderson and John Rowell found that an electric current will flow through that insulator even with no voltage difference between the superconductors. The physical process involves quantum-mechanical tunneling, as in the case of the Esaki diode, but until Josephson's understanding of the process, physicists had assumed that the electron current in a superconductor could not tunnel. When the voltage difference between the superconductors is greater than zero but less than a characteristic value of several millivolts, the junction behaves as an insulator, with zero current flow. Then, above the characteristic voltage, it conducts again. That complex behavior allows a Josephson junction to amplify power—the key requirement for a computer's logic gates. Two Josephson junctions can be put in parallel to exploit another quantum-mechanical effect, making a "SQUID," or Superconducting Quantum Interference Device. All the applications of Josephson junctions to computer circuitry considered so far involve either individual junctions or SQUIDS.

Superconducting logic circuits are microengineering pushed very far indeed. When Josephson junctions are built into circuits with conductors 2.5 microns wide, they can switch from the logic 0 to 1 or back again in about 40 trillionths of a second—or about the time it takes a light ray to cross a fingernail.

That technology is being pushed very hard in both the United States and Japan. IBM has been developing Josephson-junction technology since the early 1970s. A conceptual study by another U.S. manufacturer outlined a superconducting computer that would occupy less volume than a typewriter, yet be able to store several hundred billion words and operate 100 times as fast as a CRAY-1 or CDC's latest model, the CYBER 205. Fujitsu's Supercomputer Project is one part of a six-pronged effort launched in

1982 by MITI, an effort that includes all six Japanese computer manufacturers: Fujitsu, Hitachi, NEC, Oki Electric, Mitsubishi and Toshiba. Its goal is a computer able to do 10 billion calculations per second. Its practical applications will include the realistic three-dimensional simulation of airflows in transsonic flight down to the molecular level, and the total CAD and simulation of the most complex robots. MITI expects the Supercomputer Project to pay off with a working machine by the beginning of the 1990s.

THE ULTRASONIC MICROSCOPE AND LASER

The scientists and engineers who do microengineering constantly hunger for more precise tools to visualize and to shape the micro-world. Waves of light and waves of sound are made to serve as accurate tools for two of the exotic technologies of microengineering: the ultrasonic microscope and the laser. To inspect the invisible, deep layers of printed circuitry in today's multilayer semiconductor chips, Professor Calvin Quate of Stanford University's Applied Physics Department developed the ultrasonic microscope. A sound wave with a frequency as high as that of a microwave, about a billion cycles per second, is sent down into the semiconductor, and its echo is received. Hitachi, Olympus and other manufacturers quickly picked up the idea. Photographs of the deep interior of a semiconductor chip taken by an ultrasonic microscope detail the printed-circuit layer as clearly as a surface viewed through an ordinary optical microscope. The technique is also used for detecting interior cracks in ceramics, and for mapping the boundaries of grains in magnetic materials without having to destroy the specimen by slicing and etching.

Communicating and recording by light waves are high-growth areas of microengineering. Both were made possible by the discovery of the laser principle three decades ago in the United States. To make a laser one needs a system, molecular or larger in size, in which a long-lived energy state lies above a lower, stable energy state. The upper is called an excited state, and the lower the ground state. For example, one can take a ball and a shallow bowl and place them on a pedestal above the floor. The ball can remain in the bowl (the excited state) or it can fall to the floor (the ground state). Now imagine many such pedestal/bowl/ball combinations, all in the same room. And imagine further, for an accurate analogy to the atomic-scale laser systems, that each time a ball falls from a bowl to the floor a musical tone of a certain precise frequency is generated, and that a tone of just that frequency makes all the pedestals vibrate like tuning forks, shaking the bowls so that other balls may fall out. If one first goes through the room lifting each ball from the floor into a bowl, and then shakes just one ball loose from its bowl, very quickly a chain reaction can build up, resulting in a loud musical note at the one precise frequency, and ending up with all the balls on the floor.

In an actual laser, one loads energy into the upper, excited state ("pumps it up") by some external means, and the release of energy as the molecules fall to the ground state generates light at a precise frequency. By now a great many laser systems have been discovered. Many of them are molecular, with frequencies corresponding to those of infrared light, but the laser action has been made to occur even with free electrons in a magnetic field, and at frequencies everywhere from those of microwaves to those of visible light.

Laser light can be focused to a tiny spot because it is light of only one frequency, whereas white light consists of a range of frequencies that focus slightly differently as they go through a lens. Hitachi, Pioneer and Olympus are all particularly active in the microengineering of recording systems using laser light. Hitachi has developed a simple, inexpensive system excellent for the permanent recording of archival information. The spot of laser light is turned on and off as a cheap plastic disk spins. The system codes digital information on the disk just by burning tiny cavities inside it. In the United States, the same principle is used in the laser printer, an expensive but very powerful machine that can print a complete page of typing on ordinary paper in a fraction of a second.

Perhaps the most tantalizing of the consumer products based on the laser is the videodisk player. Its fundamental advantage over any magnetic recording system is that its recorded information is embedded in a protected layer deep in a transparent plastic disk. By contrast, the magnetic field of a recording head can be localized only at the head itself, so that the magnetic particles it aligns must be exposed on the vulnerable surface of a disk or tape. In the videodisk player the laser is used nondestructively, to shine light on a precise spot, from which it is reflected or transmitted to a detector. The potential is great, because the information is in digital form; because the quality of the recording is excellent and does not deteriorate with repeated use; and because disks can simply be stamped out mechanically in a single "shot" by automated machinery. Eventually they should be cheap, in contrast to prerecorded VCR cassettes, which must be made by the time-consuming process of running each tape past a recording head. Unfortunately, the market for videodisks has built up very slowly, and as a result the main American company involved in videodisks, IBM, has dropped out.

The Pioneer company had been doing research on videodisk technology for more than ten years when Senior Managing Director Teruhiko Isobe found that MCA (Music Corporation of America) in Torrance, California, was doing the same thing. Meanwhile, Phillips, in Holland, was far along with a similar system. Isobe felt that MCA had a better system than Pioneer for making the disks, but MCA was a laboratory rather than a production facility. Both companies contributed patents to a joint venture, Universal Pioneer, which they set up in partnership with IBM. IBM's contribution was its strength in productivity and quality control, important because Universal Pioneer planned to manufacture disks in the U.S.A. and in Europe.

Isobe estimated that the final price of a videodisk, given a big enough market, would be no more than twice the price of an audiodisk of equal duration. The actual manufacturing cost is low, but the cost of program rights (for example, to copy feature films) is equal to it when the total market is small. When the videodisk industry is mature, the disks for a motion picture should cost no more than about 20 percent of the price of the corresponding prerecorded videocassette. And the quality (resolution) of a videodisk is about 40 percent better than that of the best videocassette. But Isobe pointed out that a prime feature of the videodisk system, the ability to go to and pick out a particular frame of a movie in seconds, is largely wasted when the disk simply stores entertainment material. Disks for education, by contrast, could use that feature effectively. A single disk could store pictures of 54,000 works of art, with 30 seconds of music or voice in high-quality stereo for each picture. A few disks could store the contents of an entire dictionary.

Pioneer stayed with the videodisk project even though the initial marketing errors kept the market from rapid growth. The company was spending more than 10 percent of its videodisk revenues on research and development in 1982 even as IBM was bailing out of the joint venture. John Opel, Chairman of the Board and Chief Operating Officer of IBM, explained that IBM had been interested in videodisk technology not for entertainment but for its huge potential as an interactive educational tool. A student at a console could follow a structured-learning course based on a videodisk, and because of the system's ability to go to any frame of image and sound track within a few seconds, the system could respond flexibly and individually to his interests and to his answers to test questions. IBM's hope had been that the entertainment market would bring economies of scale to the production of videodisks and their players, and that IBM could then exploit the resulting maturity of the videodisk industry to produce low-cost instructional systems and disks. The maturity didn't come soon enough to satisfy IBM's needs, so the company pulled out of Universal Pioneer. IBM retained its videodisk patent portfolio and is pushing the technology vigorously within its own development laboratories, while already using it intensively for instruction.

The technical promise of the videodisk is excellent, and the second generation of the disk player will be far smaller than the first—hardly larger than the 12-inch disk itself. The key development that will bring it about was made in 1981 by the Olympus Optical Company. It is a miracle of microengineering, a videodisk pickup head that incorporates a semiconductor laser instead of the large, clumsy gas laser of the early models. It is called TAOHS, and is about the size of a 35mm film cassette. Two of the Olympus factories produce the TAOHS head exclusively. A videodisk player using TAOHS has been marketed by Olympus since 1982 for digital music recordings of extremely high quality.

Pioneer, Olympus and other Japanese firms are actively researching the ultimate in videodisk technology, a system that can record and play back, then be erased and record again. Pioneer is investigating at least five different ways of reaching that goal; one of the most elegant is to use a focused spot of laser light to heat a compound of cobalt and the rare element gadolinium, in the presence of an external magnetic field. When the tiny spot cools again it has been permanently magnetized, and as a result changes the polarization of light focused on it by a TAOHS head used for readout. The binary 1 corresponds to one polarization, the 0 to its opposite. To erase, the magnetic field is turned off and the laser reheats each spot that needs erasure. Isobe expects that his erasable videodisk will store 150 million bits of information—more than can be stored on 150 5-inch magnetic floppy disks.

FIBER OPTICS

The semiconductor laser is essential not only in advanced videodisk recording but in the most significant communications development of the last decade: fiber optics. As we become a more and more information-intensive society—exchanging data streams between computers of ever-greater capacity, tapping into central data banks with our own personal computers, conducting telephone conference calls over thousands of miles—we press harder on the limits of our established long-lines telephone system. The limit of a communication channel is the number of bits per second that can be transmitted, and that is set by the range of frequencies (the band width) that can travel through the channel. For that reason, designers have pushed toward ever-higher frequencies for communications; but they have reached a point of diminishing returns. In underground cables at high frequencies, signals penetrate only a thin skin of conductor, so resistance and losses are high. In the atmosphere, high-frequency microwaves are absorbed strongly by water vapor. The solution is to leap to a still higher frequency range—that of light itself. But as the atmosphere is often obscured by fog and rain, light-wave communications must travel a protected pathway, a thin, very pure glass fiber.

Glass has been an important material for humankind since its discovery nearly four thousand years ago, yet even now it remains one of the least understood of our useful materials. Michael Faraday showed in 1830 that glass is a solution, not a compound, and it behaves in many ways like a supercooled liquid. Its structure is not crystalline, as is that of many transparent solids, and for many years it was thought to have no structure at all. But modern studies using electron microscopes have shown that it has a microstructure, forming rough hexagons about a thousand atoms across. In 1966, Charles K. Kao, a communications engineer then working for Standard Telecommunications, an English branch of ITT, proved theoretically that silica (silicon dioxide), the basic ingredient of glass, could transmit light with almost no absorption if the silica could be made pure enough. More than a decade of research at Bell Laboratories and Owens-Corning was needed before glass fibers of the necessary purity could be produced economically.

In the most widely used method, MCVD (Modified Chemical Vapor Deposition), a silicon gas and carefully measured trace compounds ("dopants") are fed by computer control into a hollow glass tube about the length and diameter of a broomstick. The flame of a torch moves up and down the tube, and as the gases form particles they drift downstream, settle on the inner surface of the glass and are sintered into thin layers of perfect silica glass as the torch goes by. After some fifty passes of the torch, the temperature is raised to a level at which the tube collapses inward, to form a solid "preform." Then, on a machine several stories high, the preform is slowly lowered into a furnace. Out of the bottom comes a glass fiber only a tenth of a millimeter in diameter—eventually 9 miles of it from a single preform. Jacketed in plastic and wound on a spool, it becomes a pathway for light. The original glass tube is now a thin coating on the central fiber, which reduces losses by keeping the central fiber from contact with any lossy material. Such fibers lose only 25 percent of the light they carry per kilometer of length. Amplifiers ("repeaters" in telephone lingo) are needed only every 6 kilometers, instead of every 1.5 as for microwave cables.* A Bell System standard glass-fiber bundle, made into a single cable that can be pulled through conduits, buried underground or payed out along the floor of the sea, uses only one fortieth as much space in a conduit as the microwave cable it replaces. That, rather than cost, is the essential advantage of fiber optics for downtown city areas: replacing wire lines with optical fibers will increase the data flow 40 times.

Producing ultrapure glass fibers was only one problem that had to be solved before fiber-optics communication systems could be realized. Semiconductor amplifiers were developed to reboost the pulses of light coming at a rate of 44.7 million pulses per second, and the problem of joining fiber cables together was solved. The price of a fiber-optics transmission system is decreasing rapidly. Development is very active in Japan and the United States. Japanese firms dominate the market outside the Atlantic area, having installed commercial systems in Argentina, Brazil, Taiwan and Australia. The market worldwide for optical-fiber components grew at 45 percent per year in the early 1980s. In terms of advanced technology, Hitachi has installed systems with two cores in a single fiber, to handle two-way transmission. Hiroshi Yamada, Director

---

* A microwave cable is a metal tube through which high-frequency radio waves can be sent with relatively low losses. Those radio waves carry far more information—for example, telephone channels—than a bundle of wires having the same total diameter.

of the Fujitsu Laboratories, feels that Fujitsu is ahead of Bell and Western Electric, especially for light of 1.5-micron wavelength, which is the wavelength for minimum absorption. Dr. Michiharu Nakamura of Hitachi assesses the fiber-optics technology of Japan's NTT as the best in the world, and expects that by the end of this decade Japanese homes as well as companies will be fully "wired" by an optical-fiber system.

As for costs, Yamada of Fujitsu predicts that the cost of fiber-optics lines for local service will eventually equal the cost of conventional copper-wire lines, and that for long lines the transmission cost over fibers will be only a tenth as much as for today's long-distance service. If that translates into a substantial reduction in long-distance rates, the fiber-optics revolution in combination with the telephone "modem" (modulator-demodulator) and the personal computer will accelerate the replacement of the letter-envelope-stamp mail system by instantaneous electronic mail. And it will certainly complete the "wiring" of the average household to central data banks like The Source. It is also likely to make teleconferencing (multiperson conferencing using video and voice links) somewhat more attractive.*

The first long-distance fiber link in the United States, over the Boston–Washington urban corridor, became the subject of an acrimonious display of protectionism. AT&T opened bidding for it in 1981, and Fujitsu, Ltd., was the low bidder. But information about the low bid was leaked to Congress, Deputy Secretary of Defense Frank Carlucci wrote a letter to AT&T warning that repairs would be extremely difficult in wartime if a foreign supplier was chosen, and AT&T awarded the the contract to its own subsidiary, Western Electric. Fujitsu struck back with a lawsuit.

THE PORTABLE COMPUTER
Pressures are increasing for the development of a truly portable computer, something light and small that can be carried in a briefcase and used by the business traveler for writing reports and editing data in flight or in a hotel room at night, and then connected to the home-office data bank by telephone modem for a two-way exchange of information. That seems at first like a natural stage in the

---

* There are substantial psychological barriers, however, to teleconferencing that have no counterparts in composing messages at a keyboard or accessing data banks.

development of computers; but the transition is proving difficult because of technical problems that no one has found an inexpensive way around. Those problems, thanks to the chip revolution, are not in the fast memory or central processor. They are in the display and in long-term, power-off storage of data.

The ordinary cathode-ray tube (CRT, or oscilloscope) has been a practical, usable device for well over half a century, and is still being improved. But it is bulky, and fragile because it is a glass jug with a vacuum inside it. For a tabletop computer it makes a good, inexpensive display, but for a portable a flat screen is needed. Hiroshi Yamada and his associates at Fujitsu, Ltd., summed up the possibilities of the familiar liquid-crystal display (LCD) used in watches and calculators. It requires very little power, but the contrast between the display and its background is not very good, and the LCD becomes expensive when made in large sizes; it requires as many electrical drivers as there are dots of display to be driven. Both Hitachi and Toshiba were making small black-and white TVs with LCD displays by the early 1980s, but a computer display would have to be much larger.

For years researchers have looked for a chemical that would provide a good electrochromic reaction—would change color when an electric field is applied; no candidate so far has been good enough to be put in a product to be marketed. A very old effect, the glow of a plasma made by ions and electrons in a low-pressure gas, is being studied by NHK and other companies looking toward a large-area flat-screen TV set. Plasma displays are used for some clocks and for aviation electronics. Minoru Morita of Matsushita Electric envisions low-power, high-quality flat displays in full color by the late 1980s or early 1990s. Dr. Lewis Branscomb, Chief Scientist of IBM, told me of two flat-screen displays being explored by his company. IBM is working on a high-luminosity, high-resolution form of plasma display that could present two pages of text easily readable from all angles by a dozen people at a conference table in a brightly lit room. But a plasma display requires too much power for a portable, battery-operated computer.

Dr. Branscomb described an electrochromic display being developed at IBM's Hursley laboratories in England. It consists of a high-density memory chip about 1 inch square, with a layer of transistors on top of it, one transistor per memory cell. Each transistor drives one tiny square of conductor, all the squares forming a flat surface that is the lower side of a wafer-thin electrochromic cell. When a given memory cell is addressed by a computer, its transistor turns on, and in about one second a mirror-bright square of metal electroplates out of the liquid onto the upper surface of the cell. When the signal is withdrawn, that square of mirror dissolves again into the liquid. The chip draws very little power, and the display can be made readable with a light source and a magnifying optical system. Because it can be packaged on one chip, that system could ultimately be inexpensive. But of all the new possibilities, only electroluminescence, the glow of certain compounds like zinc sulfide under electron bombardment, is usable so far for a lightweight, high-resolution computer display screen—and such screens are very expensive, costing about 15 times as much as a CRT.

For the working ("fast") memory of a computer, about 400,000 to 1 million bits of information are adequate. Microchip memories could handle that by the late 1970s, which is what made tabletop personal computers possible at that time. But to store the programs and data that a traveler may need on a trip, a minimum of 2 million bits of long-term memory are needed, equal to the capacity of two or three 5-inch floppy disks. And for a portable computer, little or no power can be drawn to maintain that information. That rules out the ordinary N-MOS memory chips, which draw enough power to drain a battery within a few hours. In the long run, the most cost-effective lightweight solution to the long-term, or "nonvolatile," memory problem may turn out to be 256k C-MOS chips, with one of the new-generation 3-inch floppy disks or a microcassette used to transfer programs and data to and from that nonvolatile memory. An elegant but expensive alternative for long-term storage, available since the early 1980s, is the magnetic-bubble memory.

The magnetic-bubble memory consists of semiconductors containing magnetic elements in which it is possible to establish a checkerboard pattern of squares, with the magnetic field of all atoms in a square pointing in the same way. Moreover, one can form the material so that each square has just two stable conditions, magnetic field up or field down—and will remain in either condition indefinitely, requiring no power. But to go from that realization to a salable commercial product took a heavy investment in research and development. Intel remains a leading manufacturer of bubble memories, but the price of production did not come down as rapidly as had been hoped, the market therefore remained much smaller than expected, and one by one the early U.S. manufacturers dropped out. Rockwell International gave up early in 1981, Texas Instruments stopped production a few months later after spending $50 million, and National Semiconductor closed its bubble-memory production facilities shortly afterward.

But Japanese companies showed greater endurance. Hitachi and Fujitsu, slower than Intel to produce a bubble memory of a million-bit (1-megabit) capacity, installed new production facilities for that later-generation chip and went into competition with Intel across the board for sales in the United States. The Japanese firms worked from a strong base in the home market, where they had been producing 256k bubble memories for automated machine tools and for computers.

Recognizing the great potential of the portable-computer market, several firms in the United States and Japan made computers with carrying handles. To keep prices down, they stayed with conventional display and memory technology—CRTs and 5-inch floppy disks; but the weight of the machines, usually 17 to 23 pounds, made them unsuitable as portable computers.

The alternative to building the heavy "anything with a handle is portable" computer of the early 1980s was to press the technology to its limits, ignoring expense. GRID Systems took that approach. GRID was started in 1980 by John Ellenby, and its potential market was the larger U.S. corporations, which Ellenby estimated had by then installed $400 billion in large mainframe data-processing hardware and supporting facilities. Ellenby's plan was to make the data base for every such firm available to every executive, whether in the office, at home or traveling. An essential component in the plan was the "Compass" portable computer and its "Navigator" software system.

The Compass weighs just over 9 pounds, and packs an incredible capability within a magnesium case measuring $2 \times 11 \times 15$ inches: an Intel 8086 16-bit microprocessor, an 8087 high-speed algebra-calculating chip, 2 megabits of fast memory and—a costlier item—another 2 megabits of nonvolatile magnetic-bubble memory. It has built-in modems for sending and receiving over telephone lines electronic mail, programs and data at both the standard Bell System rates, 300 and 1,200 bits per second. It has plugs for cable connections to external disks and other computers. It can run on either 110 or 220 volts. For its display, the GRID engineers bought something never before used except in military equipment: an electroluminescent panel. The panel, made by Sharp Electronics in Japan, is only millimeters thick, but provides about the same resolution as a CRT over a 4 × 5-inch area. And as might be expected for a machine with such high capability, the price of the Compass is correspondingly stratospheric—about $1,000 a pound. Japanese firms see an equivalent of the Compass available (in a plastic rather than a magnesium case!) by about 1985 for approximately a quarter of the price.

COMPUTER LANGUAGE AND "TERMINAL PHOBIA"

These accomplishments of microengineering are impressive. But what about all those people with "terminal phobia": people who are uncomfortable with technological artifacts, who feel more and more alienated as they hear their friends and their children comparing the fine points of computers and computer languages?

The most exciting trend of the 1980s in microengineering is the trend toward making computers easier to communicate with. Grace Hopper set us on that course in the 1940s with her first compiler for computer programs, and Dr. John Kemeny of Dartmouth made a giant step in the early 1960s when he invented BASIC, an easy, English-like language of a very high level.

It was a stroke of genius in Kemeny's design that BASIC was qualitatively different from other high-level languages. It was an "interpreter," not a "compiler." Other computer languages—PASCAL, FORTRAN, COBOL, PL-1 or any of the rest—take the set of instructions one writes in the language and make up a much larger set of machine-language instructions (called the object code) to do what you ask. By contrast, Kemeny's BASIC is a whole set of minicompilers, one for each instruction you write. As you write your program, you can run it at any stage, and check for errors instruction by instruction. However, BASIC runs slowly compared with programs that are compiled as a single entity. But by the early 1980s, programming ingenuity solved the problem. One could write a program in BASIC, find its errors and correct them easily instruction by instruction, and then when it ran perfectly turn it over to a special supercompiler to be rewritten into an efficient object code that would run about six times as fast as the sequence of BASIC minicodes.

The next step was a program that would write programs. That

art is still young, but there are superprograms available now, like the "COBOL Program Generator" from David R. Black and Associates, or "The Last One," from DJ "AI," Ltd., in England, that ask you a sequence of "What is it you'd like to do?" questions and generate a complete bug-free object code from your responses. As to their value, note that computer companies long ago worked out as a general rule that an experienced, hardworking programmer, writing without the help of such a superprogram, can generate only about two lines of error-free program per day of work.

As the programs that write programs are perfected and used increasingly by programmers, it will become easier and cheaper for customers to buy software that is much more "friendly." Until the late 1970s, only professional programmers, scientists and engineers, all willing to devote substantial time to programming, could develop software suited directly to the problems they needed to solve. In the estimation of Dr. Lewis Branscomb of IBM, the 1980s will be increasingly an era in which technically educated people without special programming experience will be able to write their own programs. The 1990s, in his opinion, will be a time in which people even without any technical background will find it easy and comfortable to use computers.

For those people whose terminal phobia prevents their doing any program writing at all, the most attractive operating system languages of the 1980s are likely to be Xerox' SMALLTALK and ADA, which was developed by the Defense Department and is similar in concept. SMALLTALK is sophisticated enough to shield the user from operating systems, text editors, filing systems or programming languages. Instead it gives the user all those functions simultaneously, and allows him to move from one to another without consciously crossing any boundaries. SMALLTALK was developed at Xerox' Palo Alto research laboratory over a ten-year period. It recognizes that humans don't think in a logical straight line, one item at a time. Instead we associate items in parallel; one thought suggests another. A desk with several papers and pictures on it at once, all visible, isn't a sign of confusion but an efficient way to make many sources of information simultaneously available to our associative, parallel-thinking minds. SMALLTALK presents an image like many pieces of paper, partially overlapping. Each piece, called a "window," can have graphics and text intermingled without restriction. When you want to use a window, you "pull it

80

out" and concentrate on it while the edges of the others remain visible and identifiable. The Apple Corporation's "Lisa" computer, introduced in 1983, used that principle and further eased the task of users by providing a controller called a "mouse." To move the active dot on the screen, called the cursor, to any location, it was necessary only to move the mouse, without the need of the keyboard.

We can expect the Japanese to develop new and excellent associative programs, because their written language and their cultural traditions make them very good at pattern recognition. As Makoto Kikuchi, Director of Research for SONY, points out, Japanese children learn the geography of their home cities through the patterns made by landmarks. Most streets in Japanese cities did not even have names until after World War II. And each of the 2,000 or so Kanji characters (Chinese ideographic characters as they are used in Japan), which any high school–educated Japanese must know in order even to read a newspaper, expresses not one but many meanings, depending on its association with other characters. Kikuchi points out that as a result, Japanese children score extremely high on pattern-recognition tests. He describes English as a "linear, specific, binary language well suited to the sciences," and thinks it natural that computers of the kind we have today were developed in the English-speaking world. In contrast, he notes that Japanese has no well-defined "yes" or "no" but communicates by associations, in a manner that works only because the Japanese are a homogeneous people all of whom understand the same thing by a given reference.

Building on the native Japanese talent for pattern recognition, MITI and five Japanese firms (Fujitsu, Hitachi, Mitsubishi, NEC and Toshiba) spent about $100 million in a ten-year program to develop electronic recognition of graphic patterns and human speech. The target for optical character reading was met: machine reading of 1,000 different printed Kanji characters with an accuracy of 95 percent. For the easier task of recognizing English text, Toshiba now sells as a result of that program an optical character reader capable of reading any of four different type fonts, with near 100-percent accuracy. But Kanji remains a difficult medium for electronic communication. Dr. Michiyuki Uenohara of NEC feels that today's binary computers are not at all well suited to Kanji, although computers that communicate in Kanji do exist. As for keyboard input, a fast professional typist working at a Kanji typewriter can communicate only one third as fast as a typist working in English. But Dr. Michiharu Nakamura of the Hitachi Research Laboratories pointed out that the difficulties of Kanji can even be a hidden advantage for the Japanese: young people in Japan buy English-language personal computers in order to be able to alphabetize and to use international library resources, and as a result become more fluent in English.

While Kanji remains an intractable medium for binary electronic communication, spoken Japanese is a relatively easy language for a computer to recognize. The language has fewer different sounds (phonemes) than does English, and it is not a subtly inflected, accented language. One can speak a Japanese phrase in a number of different ways, with quite different accentings, and still be understood. That has made it possible for Japanese firms like NEC, Mitsubishi and Matsushita Graphic to go quite far with electronic systems for the recognition of speech. NEC met the goals of the ten-year MITI program by developing a system that could recognize a 200-word vocabulary spoken by anyone, with 98-percent accuracy. Japanese prefer oral to written communication, so there is a strong drive to develop inexpensive office equipment with voice-recognition capability for Japan's home market. The technology developed there will certainly be used in export equipment, but Japan will have stiff competition. Lewis Branscomb of IBM believes his company is the world leader in the recognition of continuous, uninterrupted speech, though IBM recently backed off to an emphasis on the recognition of single words in order to develop marketable products quickly.

According to the researchers working on electronic speech recognition, it may be the early 1990s before a computer could recognize spoken language well enough to be able to serve an executive as an efficient secretary.

Graphics recognition, speech recognition and three-dimensional computer-aided design will call on a great deal of what Branscomb calls "sheer raw compute-power." For those tasks, the exotic Supercomputer in which Japan is now investing heavily may be adequate. But all such machines are linear and binary, and therefore cannot "think" in the same way as our brains.

The most ambitious microengineering project of all addresses that problem: it is Japan's "Fifth-Generation" Computer Project, which began in 1982 as a three-year first phase of development. Though it will probably use Josephson-junction and high-speed gallium arsenide technologies, like its more conventional cousin the Supercomputer, it will process nonnumerical information like pictures and graphs directly. The Japanese who are active in it, notably Professor Tohru Moto-oka of the University of Tokyo, have studied closely the twenty-five years of investigation that U.S. researchers put into "artificial intelligence." Their goal is not just a computer that is faster and remembers more, but a machine that will think like a human being: a "knowledge processor" rather than simply a "data processor." It will be capable of judgment, weighing qualitative as well as numerical information. The machine will be compact enough to be located in an office, and it will interact with human beings just as another human would, by normal speech with a vocabulary of 10,000 words, supplemented by graphics output and the ability to recognize and understand handwriting and scrawled sketches. MITI has established an independent research association for the Fifth-Generation Project, and its staff of several hundred is being drawn from all the top Japanese computer firms. Its budget may run to $400 million, and the project will take a minimum of ten years and perhaps as many as twenty. What is the main driving force behind the Fifth-Generation Project? Americans and Japanese offer very different opinions: Professor Moto-oka says, "This project is the space shuttle in the world of knowledge. We Japanese must show our creativity to foreign countries." Darrell Whitten of the Bache Company says the main Japanese motive behind the Fifth-Generation Project "is not just to coexist with IBM, but to beat IBM." If the Japanese succeed with the new family of machines, they will render obsolete the linear, binary information technology that has been developed by IBM and by other Western firms over the past four decades.

CHAPTER 3

ROBOTS—THE NEW BREED

Robots in one Japanese factory now manufacture replicas of the motors and gears that are their own internal organs. Sometime in the next decade, machines that are still more advanced will achieve the beginning of "robotic life." They will complete the assembly of the first robot wholly constructed and assembled by others identical to it.

THE EVOLUTION OF ROBOTS

Robotic life is the lineal descendant of the numerical-control (NC) machines that manufactured components of airplanes in World War II. Those primitive robots, some of them as big as a house, milled, drilled and bent aluminum and steel following directions on punched paper tapes.

Since then, robots have evolved from those specialized ancestors capable of working only in very narrow ecological niches to general-purpose, adaptive, intelligent species able to work in a wide variety of environments. Microengineering gave them intelligence in the 1960s, when the first inexpensive minicomputers were born. The earliest companies to seize upon that new technology to manufacture general-purpose robots guided by complicated programs were Unimation, in Danbury, Connecticut, and Cincinnati Milacron in Lebanon, Ohio. The hydraulically driven "Unimate" robot produced by the Danbury firm was developed from a project at Stanford University's Artificial Intelligence Laboratory. Unimates are found today in heavy-machinery factories throughout the industrialized world, doing the hot, heavy, repetitive jobs that are most boring and most dangerous for men. The largest, weighing more than 2 tons, handles a load of 450 pounds in a gripper capable of all three rotations of the human wrist and forearm; it has a telescoping upper arm that can swing up or down, pivot left and right, and extend in and out. The total motion available is six degrees of freedom, or as the industry says "six-axis control." While ideal for welding and for manipulating engine blocks and heavy, white-hot ladles of molten metal for casting, such machines are not highly precise: they can reproduce a sequence of movements programmed for them (usually by a floppy disk in a minicomputer) only to a precision of about 2 millimeters—equal to the thickness of a phonograph record.

The Nissan Motor Company in Kamakura has dozens of Unimates, all manufactured under license by Kawasaki Heavy Industries, welding Datsuns on the assembly lines. By 1982, about 97 percent of the welds on Datsuns were being made by robots. When a robot fails, a human worker is called in to take the robot's place until the shift ends.

Machines much smaller than welders, in Unimation's "Puma" series, add a seventh degree of freedom, the elbow, and can pick and place small components of a few ounces' weight with a precision equal to the thickness of a sheet of paper. Hitachi and other Japanese companies manufacture intermediate-size robots suitable for painting; they move a spray head in a reproducible straight line between start and end points that can be "taught" to them by a human worker. General Electric, in entering the robot market, chose to begin by importing Hitachi robots for sale under its GE trademark.

The secret of making highly accurate but inexpensive robots is to provide motion by a simple, imprecise machine, but to guide the robot by a sensor that can measure position accurately. Such companies as International Robomation/Intelligence use that principle now in a modern generation of robots with shoulder, elbow, arm, wrist and hand motions that can position loads of 50 pounds with great accuracy. The robots measure the difference between where the load is and where it should be—just as we do by eye. Such machines, guided by a minicomputer, can operate in atmospheres with explosive dust or vapor because they are driven by air motors rather than by electricity. They cost no more than an average car.

THE ECONOMICS OF ROBOTS

The money that can be saved by using robots on production lines is a survival issue in our competitive world. Panasonic found that the incidence of defective products from its color-television factory in Japan dropped to a fiftieth of the previous rate with the introduction of robots into assembly work. Five component-insertion robots in the Panasonic factory, tended by 4 human workers, did the work of 44 people whose salaries and benefits would have cost nearly $800,000 per year—enough to pay for the robots in less than six months. No wonder that Japan's demand for and output of industrial robots grew by 85 percent in 1980 alone.

Worldwide, the market for robots is expanding a little more slowly. Sales multiplied by 6 times in just the four years 1978–1982, for a 55-percent annual increase. On that basis, the industry's standard estimate for growth during the 1980s, 35 percent per year, may be on the low side, but it is still very nearly an explosive rate. The world total of robots will multiply by 20 times in just one decade, and yield a market of more than $3 billion per year by 1990.

Both in the total of installed capacity and in output, Japan leads the world in robots. By 1982, Japan had 14,000 robots (using the more stringent American rather than the looser Japanese definition of a robot) on its production lines, while the United States had only 5,000. Japan had by then captured 45 percent of the world market for robots. Very large U.S. firms like IBM, which had previously made robots only for their own assembly lines, began selling them to the open market only in 1982.

The robots built in the first decades of the robotics age were designed mainly for simple, highly repetitive tasks requiring great strength. The emphasis now is on much more complex motions, and on the ability to store a number of different sequences in memory. The Unimation/Kawasaki welding robots at Nissan can follow any one of eight different geometric patterns, corresponding to different models of cars and trucks. Many of the new-generation robots are of modest size and very high precision, because those are the machines needed to make the products of microengineering, like VLSI chips, computers and VCRs.

Industry needs robots for production lines, and has little need for mobile robots built in the semblance of man. Developments along that path are being left mainly to hobbyists, many of them in their teens and 20s. They took to personal computers enthusiastically when that concept was new, and now they're taking to robotics in the same way. If anyone succeeds in making a usable manlike personal-servant robot in the next decade, that someone will probably be a hobbyist, because the professional industrial designers are busy working on machines intended for service on assembly lines.

The task of making a robot that can move about like a man is very difficult; the early attempts, while highly ingenious, are still primitive. The "Avatar," a robot made by a hobbyist named Charles Balmer, Jr., took several years to build out of scrounged parts, and another year of programming before it could go about its few simple tasks: finding its recharger "oasis" when its battery began to run down, responding to simple one-word voice commands, roaming a house without bumping into walls and fetching its master's toolbox or slippers. Avatar had a vocabulary of about 400 words, and could recognize 30 words spoken by its builder. Sometime in the next century, when companies try seriously to produce a robot servant with humanlike responses and behavior, the most difficult task will be programming; computers operate differently from humans' associative, parallel thinking.

TEACHING ROBOTS

Because of the mechanical strength of typical industrial robots, and their substantial cost, production firms find it difficult to train programmers on them. Either an expensive production-line machine is tied up while a programmer is learning, or something expensive may get broken when the robot follows a faulty program. That has given an enterprising small company an exploitable market niche: manufacturing inexpensive, not-too-powerful robots to be used for practice by industrial programmers. The company, a Silicon Valley start-up that was formed in 1981, is called Microbot. Its most popular product, the "Teach-Mover," is a five-axis machine that can swing left or right, move its arm up or down from a shoulder, bend its elbow, rotate its wrist both up and down and in twisting motion and grip anything up to 3 inches across that weighs a pound or less. The programmer uses a push-button device about the size of a pocket calculator to position the Teach-Mover correctly, then pushes a "record" button to write all five coordinates into a memory. The little machine works to an accuracy of better than a millimeter within a sphere of about 1-meter diameter, and its memory can hold 53 different sequential positions. The Teach-Mover costs less than a typical personal computer. Microbot's even less expensive "Mini-Mover 5" gets its instructions from an Apple II, a TRS-80 or any other computer of that class. Microbot's small robots are being bought in increasing numbers not only for teaching but for the same sort of production assembly jobs that are done by their more expensive cousins.

None of these useful, productive robots has the ability to move itself from place to place; but for the first time, robots with their own mobility are beginning to tackle one of the most dangerous jobs in the construction industry: boring tunnels. Tunneling through solid rock is relatively accident-free, though slow. But tunneling through mixed soil and mud, a much more common challenge, is dangerous because the tunnel must be pressurized. Too much pressure and the tunnel may blow out. Too little, and it may collapse. Regardless, the men working behind the drilling head must enter their working area through airlocks, and in their off hours may suddenly be seized by the bends, like deep-sea divers. Mitsubishi Heavy Industries has developed a robot tunnel borer, 12 feet in diameter and 17 long, that has a rotary head to cut and chew its way through the ground. The material excavated is conveyed back through the cutter to an automated miniature train that carries it away. In 1981, a Japanese construction firm, Ohbayashi-OAC, leased that "Mole" and brought it to San Francisco, where it was used successfully to bore a hole about 1 kilometer long. The completed tunnel was lined with concrete, and became a pipeline to convey San Francisco sewage to a new treatment plant, reducing the city's pollution of San Francisco Bay by 90 percent.*

ROBOTS AND INDUSTRY
The ultimate purpose of industrial robotics is to lower production costs while raising quality. But as long as the finished product must consist of a large number of subassemblies, there is a barrier to how

---

* An alternative robot tunnel borer, still untried so far, has been studied by D. L. Sims at the Los Alamos Scientific Laboratory. It is called the "Subterrene," and it melts its way through gravel and soil. As it proceeds, the heat it gives off compacts the original material of the tunnel volume into a hard glass wall. Though demanding of energy, the Subterrene is economical, because it carries out in one operation all the functions of today's tunneling systems. A small demonstration model of the Subterrene was operated more than twenty years ago.

low costs can be brought. That barrier can be eliminated only by the introduction of a fundamentally different kind of robotics: the construction of a finished product from inexpensive raw materials in a single, integrated production operation. It requires what is called a "flexible manufacturing system." The casting of metal exemplifies that concept, but the alloys needed for high-performance metal parts are unsuitable for casting.

Japan's Agency of Industrial Science and Technology (AIST), a division of MITI, has been pursuing since 1977, in a $60-million project, an exotic kind of flexible manufacturing. The aim is to demonstrate a general-purpose manufacturing system in which cutting, welding and the surface treatment of metals will all be done by high-power laser beams, and complex forming and assembly will be done by nonspecialized, adaptable machines. The entire factory is to be directed by computers.

DIRECT COMPOSITE WINDING

Machines and techniques usable for industry may well come out of that project; but in the 1970s, production experts in the United States were already beginning to sour on the time-consuming, energy-intensive, wasteful approach of cutting metal to sculpt from it a finished part. Instead, they were developing more effective machines to shape metal by forging or pressing in a single, rapid operation; "Move, don't remove" was their slogan. It seems likely that with a new kind of robotics, so far little known even in factories, we may be able to approach the goal of flexible manufacturing from common, inexpensive materials, while saving energy. The new technique is called "direct composite winding." Its leading authority and practitioner is Brandt Goldsworthy, of Goldsworthy Engineering in Torrance, California. Thin strands of a material with high tensile strength are combined with a matrix of softer material to make a complex shape with high strength in tension, bending and compression. Concrete laced with steel reinforcing rods is an old-fashioned but still very useful composite material.

The Goldsworthy factory employs about 60 people, and its front office is a modern building in a landscaped industrial park. The offices connect to an assembly area made up of two hangars with large doors leading onto the ramp of Torrance Airport. When I landed at Torrance, I was able to taxi up to the Goldsworthy factory and park within a few feet of it. Brandt himself is the picture of a man who is doing what he most enjoys.

Goldsworthy's father was a mining engineer, and he was brought up in "about twenty-one different mining camps in California, Utah and Nevada." He graduated from the University of California at Berkeley as a mechanical engineer in 1935, moved to Southern California and became one of the earliest surfing fanatics. In his nonsurfing hours he worked at Douglas Aircraft.

At that time, surfers were making a transition from long, heavy koa-wood surfboards of a type originally used by the Hawaiian kings to light balsa-wood boards. The new boards were very fragile, and Goldsworthy wondered whether he could set up a company to coat them with one of the early plastics, cellulose acetate butyrate, for greater strength. No chemist, he looked up the mixture in a handbook, ordered a barrel of plastic powder and several jugs of acetone, and mixed the two ingredients together in a large earthenware crock. The result was a sticky goo, almost impossible to work with and even more difficult to get rid of.

Just then, Douglas came to him with a problem. The company had built a new airplane (the DC-5, subsequently the DB-7, forerunner of the A-26 and B-26 attack bombers of World War II). Test pilots found that the control cables of the new plane rattled in resonance exactly at its cruising speed, so that flying in a DC-5 was as loud as standing beside a riveting hammer. Brandt needed to find a material to use for the fairlead bushings the cables passed through, a material strong enough not to be cut by the cables yet resilient enough to absorb their vibration. Dozens of ideas were tried, but none of them worked. Then, in desperation, Goldsworthy made a fairlead out of ordinary commercial felt, and to give it strength dunked it into his crock of butyrate. The result worked, and Douglas then faced the problem of how to specify it for mass production. Goldsworthy made an offer: "Give me the purchase order and I'll make them on my back porch!"

Brandt is a composite expert. In his Goldsworthy Engineering plant and laboratory there are several million dollars' worth of machines built by his firm to solve an embarrassing problem the Army had discovered: it had engineered a new 155-mm artillery shell, and then had found after making many thousands of the new shells that they were too heavy to go the specified distance. Goldsworthy invented production-line machines to replace a wide band of steel in the shells by fiber glass, for lighter weight.

In the pilot-plant area at Goldsworthy Engineering there is a machine Brandt built for NASA. It is designed to fly in the Space Shuttle, and in orbit to fabricate continuously and automatically a triangular beam about a meter on each edge, out of carbon-fiber-reinforced fiber glass. Another device in the pilot plant, called a "pulltrusion" machine, fabricates a continuous fiber-glass strip out of individual threads ("roving") and polyester resin. Glass threads from a number of keg-sized bobbins come together at a roller, pass down into a tray of liquid polyester and then out again and are pulled through an accurately machined rectangular hole, the analog of the die in an aluminum-extrusion process. Then the moving ribbon passes through a heater, where the matrix is cured to hardness within seconds. Out of the end of the machine, pulled by a pair of rollers, comes a ribbon of fiber-glass strip, rectangular, with its edge dimensions accurate within thousandths of an inch.

The traditional method for making fiber-glass parts is called "hand lay-up." Sections of glass-fiber cloth are placed on a female mold and painted with epoxy by hand. As Brandt says, that method requires "eighteen girls at a long table, hand-laying the cloth and slapping goo on it with paintbrushes." Layer after layer is put on until finally the required total thickness has been built up and the whole assembly can be put into an oven to cure. That is the way fiber-glass boats are made, and most of the body pieces on the Chevrolet Corvette. Until 1967, Goldsworthy's factory worked the same way, but

I got so upset about people doing everything by hand that I switched over to engineering production machines. You could pound out a car fender with a mallet on a sandbag, but nobody does. And it's the same way with composites: you need specialized machines to make things economically. And above all, you need to design the product along with the machines.

He believes in totally automated production, and has proved his philosophy in practical, marketable products. In another corner of his plant I saw a computerized machine that was winding a test piece for a helicopter control tube, a part that had to be light, strong and 100 percent quality-controlled. A single mile-long glass fiber passed from its bobbin through a movable needle eye and onto a rotating spindle. Under the guidance of the computer, the rotation speed of the spindle and the position of the eye along the spindle were both controlled. As the test proved, the machine was building up a complete section of the fiber-glass tubing, reinforced by heavier thickness at its ends and varying along its length, entirely from one fiber that was placed within a thousandth of an inch.

That process of Goldsworthy's has been applied in a patented method now used under license in Japan to manufacture automobile and truck leaf springs for Toyota entirely from fiber glass. The spring is a solid, monolithic black object molded around steel tubes at its ends and shaped a little like a bow. It weighs 8 pounds; the steel spring it replaces weighed 40. But the fiber-glass spring absorbs road shocks better, lasts far longer without fatigue and costs much less to make. Brandt chuckled that the black color was due to a pigment added purely for psychological reasons. "If you try to sell people a spring they can see through, they won't believe it's strong enough. But something solid and black—now, that's strong!"

Goldsworthy has licensed the same process to General Motors and to his friend Dr. Kasuga of Nitto Boseki. Like most of the Goldsworthy processes, that one had to be brought to the pilot-plant and initial production stages at his own factory before his customers would believe it could work. The springs are made fully automatically at a rate of two per minute, and are cured in seconds by radio-frequency heating, the same principle that is used in microwave ovens. His firm has also made a test model of a new truck body for General Motors. It is a single fiber-glass piece made by automatic winding, with a geodesic truss of triangular reinforcement patterning the inner surfaces, like the structure of a 1940-vintage RAF Wellington bomber. Doorways are reinforced by pull-trusion strips that can be produced automatically as the main body is being wound.

Brandt Goldsworthy shares my own enthusiasm for the concept of a strong, lightweight, inexpensive airplane made entirely by automated machines using his direct-composite-winding method. And he is critical of conventional aircraft engineers, trained to design in aluminum, who approach a fiber-glass-aircraft project as if it were metal and try to simulate aluminum spars, stringers and ribs by the slow, costly method of hand lay-ups. In his view, any such plane should be designed as a composite structure from the start, and its wholly automated production machinery should be designed along with it. Changes in the aircraft shape, if necessary, should be made by software programs in a computer, not by the making up of new forms. And he concludes with a picture of aircraft production that he is sure could be reality right now:

"You can just push the button, stand back and wait until the machine hollers 'Finished.' " But for Brandt Goldsworthy, life is clearly not a process of standing back and waiting.

ROBOTIC SENSORY PERCEPTION

In the 1970s, robots were given agile, dexterous hands and fingers through the guidance of minicomputers. In the 1980s, the main thrust in the development of individual robots (as contrasted with their organization into complete manufacturing teams) is to give them the senses of sight and touch.

The optical encoder is a device that measures the angle of a shaft by counting the number of pulses of light that have come through a disk with regularly spaced holes. It is hardly a new concept, but it is inexpensive, and it is being applied widely now, to tell robotic machines exactly where their "end effectors," or robotic hands, are located. Teledyne-Gurley of Troy, New York, makes linear optical encoders that can subdivide a millimeter into 2,000 equal parts, for a resolution of half a micron. Its rotary encoders subdivide 1 degree into 400 parts, so that the lead screw in a lathe or mill can be turned just the right amount to bring an end effector to where it is needed. Brandt Goldsworthy's direct-composite-winding machines make extensive use of optical encoders to position each turn of roving exactly where it should be.

The touch sensors made by Robot Sensor Systems of Temple City, California, detect forces and torques, and feed back their information through the robot's computer brain so that it can perform delicate tasks, like putting nuts on bolts or assembling fragile parts. It can also help prevent catastrophic mistakes, like an end effector's squeezing a fragile part so hard that it breaks.

Touch is not the only sense that robots are acquiring. Robotic vision is advancing rapidly, with much of the research involving the problem of pattern recognition; almost none of the companies making vision systems shipped its first product before 1980. All of them use solid-state television cameras for eyes. The Machine Intelligence Corporation of Sunnyvale, California, specializes in two-dimensional pattern recognition. Its systems compare the camera image of a part with a master reference image stored in its memory, to recognize one particular shape among many. For example, one vision system inspects VLSI chips under high magnification to check for printed-circuit errors. Another, linked to a Unimation "Puma" robot, sorts a stream of parts passing on a belt.

Automatix, Inc., of Burlington, Massachusetts, makes a more expensive and sophisticated "Autovision" system that can examine up to six pieces every second, presented to it in a random orientation, and can recognize them according to preprogrammed characteristics like area, the length of the outer edge, the number of holes and the visual center of gravity. The industrial buyers of these systems are saved the need for programming, because the machines can be "taught" by being presented with a named object in a number of different orientations before being put to work sorting. At each orientation, the minicomputer calculates the area and other characteristics so that it can recognize the object later. Robotic Vision Systems of Melville, New York, combines with its video camera a light beam that can be pointed accurately by computer, to build up a three-dimensional picture of a complicated shape, like an engine block.

Japanese robot manufacturers were a little slower than the Americans in entering the field of robotic sensory perception, but when they did, no more than two years later, it was with giant multibillion-dollar industrial companies rather than with medium-scale firms and Silicon Valley start-ups. Fujitsu, Ltd., the multibillion-dollar computer company from which Fanuc, Ltd., spun off in the 1960s, is developing, mainly for use on its own assembly lines, a very compact robot about the size and shape of a beer keg, equipped with vision and several microprocessors. The four-axis robot uses the highly magnified visual information its camera feeds back to position its 1-kilogram load within an accuracy of 0.6 micron— about a hundredth of the diameter of a human hair.

IBM VS. THE JAPANESE

We are seeing for the first time a face-off between large Japanese robotics firms and a major U.S. company. IBM, the third-largest industrial firm in the world by my definition, began selling robots only in 1982, although it had been active in research and development for fifteen years before that, and had automated its own production lines extensively when robotics first became practical. IBM had concluded that the Japanese were well down the learning curve on the production of relatively standard robots, so it contracted with Sankyo Seiko, Ltd., in Tokyo to produce the low-cost four-axis Model-7535 robot, according to IBM's own design.

John Opel, Chairman of the Board and CEO of IBM, told me that IBM keeps within its own factory in Boca Raton, Florida, the production of its most advanced robot, the RS-1. That machine has six degrees of freedom and is hydraulically driven for great strength and precision. It has tactile sensors combined with a pattern-recognizing vision system, and its sensors monitor end-effector position, touch and sight 50 times per second, feeding back through an IBM Series 1 minicomputer to react to changing conditions. Opel and IBM's Chief Scientist, Dr. Lewis Branscomb, take special pride in the highly sophisticated programming language, AML (A Machine Language), that IBM has developed to direct its robots.

Branscomb argues that the "teach" method for programming robots is too limited and inflexible for a robot with the power and complexity of an RS-1; he feels that one can't lead the robot through every set of circumstances it must face. Also, he would like to see the day when, through AML, robots can be instructed directly from blueprints. In Branscomb's view, the real payoff on robotics should be higher output quality. He forecasts a time when automation will be so advanced that most products can be made with zero direct-labor content. His ideal for industry is "zero direct labor at the assembly line. Not to do away with jobs, but to do away with unskilled jobs, and replace them with higher-level jobs that require lots of retraining."

For tasks that do not require the strength and precision of an RS-1, IBM also markets a stripped version of AML that can be used with an IBM Personal Computer (PC) to direct a group of the simpler Model-7535 robots, for assembly, packing and loading.

THE AUTOMATED FACTORY

The second major thrust of the 1980s in robotics is the combination of a number of different robotic machines to form a complete automated factory that can operate entirely without human workers. In its most advanced form, this could become a system capable of replicating all its own components, and therefore of growing geometrically in capability.

In the late 1940s, John von Neumann, of the Institute for Advanced Study in Princeton, worked out the basic mathematics for self-replicating factory systems, often called "von Neumann machines." But only in Japan has the concept been pushed close to reality. At the beginning of this decade, the Japanese MITI planned in considerable detail a $57-million demonstration project aimed toward a totally self-replicating factory system. At present the project is dormant, the victim of budgetary constraints, though less ambitious robot factory systems are already in regular and productive operation in Japan.

For industries in which production runs are very long (several hundred thousand or more identical units), it has been cost-effective for several decades to manufacture with the help of specialized single-purpose machines; that is called "hard automation," because it does not give the flexibility of which true robotics is capable. Black & Decker in the United States uses hard automation to mass-produce a line of good-quality, inexpensive power tools for light-industry and home-workshop use. The John Deere farm-machinery factory in Waterloo, Iowa, is also highly integrated and automated and makes extensive use of computer-directed materials-conveying equipment and hard-automated production machines, together with some robotics.

But Japan is where the automated factory has been brought to its highest state of evolution. In 1982, the Yamazaki Machinery Works, Ltd., in Nagoya started operating a new "flexible manufacturing factory" which produces more than seventy different kinds of lathe and other machine-tool parts. Twelve people cover the first two work shifts at Yamazaki, Ltd., while the computer-directed production machines work the night shift with no human attendants at all. Formerly, Yamazaki's factory required 215 people for the three shifts.

The world leader in development of the fully automatic factory is also in Japan, and its leadership has given it not only a high growth rate but a consistently high rate of profit. Fanuc, Ltd., a firm of moderate size with only 950 employees, has developed and grown during its nearly thirty-year history under the leadership of one man, Dr. Engineer Seiuemon Inaba. Its products over the years incorporated some designs licensed from abroad, but the company has been overwhelmingly a developer and exporter of ideas with a consistent pattern of innovation.

Dr. Inaba's company began as a division of the Fujitsu, Ltd., computer firm, and became independent as Fujitsu-Fanuc NC (the

N and C standing for "numerical control") in 1972. It has grown into an international corporation, now called Fanuc, Ltd.*

On the basis of its products and its growth rate, one would describe Fanuc, Ltd., as a highly innovative, well-run leading firm in the numerical-control business. But it rose to a class by itself with the completion in 1980 of its Fuji Factory, in the Yamanashi Prefecture on a wooded plain just at the edge of the lower slopes of Mount Fuji. There it manufactures industrial robots, NC cutting machines and turret drills. Some of those products are manufacturing machines identical with those of the assembly line that built them— true robotic reproduction. The materials that feed the Fuji plant are brought into a multilevel "automatic warehouse." Under the direction of a central computer system, automatic unmanned carriers loaded by robots transport materials to "machining cells," each consisting of a numerically controlled cutting, shaping or drilling machine served by an industrial robot. On completion of an operation, the robot places the workpiece on a tray at the cell boundary (neatly setting each piece next to the previous one) and when the tray is nearly full signals for another unmanned carrier. When the carrier arrives, the robot turns around from its work for a moment to slide the full tray onto it. The tray is then taken to another machining cell or to another automatic warehouse, this one for parts and subassemblies. From the warehouse, the components of a complete robot or NC machine are taken by unmanned carriers to an assembly area for completion by another bank of robots.

In one of the world's more beautiful industrial sites, close to Lake Yamanaka, with the snow-capped peak of Mount Fuji filling the background, most of the time there is no human worker present to appreciate the view. The Fuji plant is organized for totally automatic operation, and routinely works five days a week, three shifts per day. Human workers in their smart yellow uniforms come in to service the machines and use computer-aided design to program for

---

* It made an agreement for mutual assistance with West Germany's Siemens A.G. in 1975, and the General Numeric Corporation in Chicago was formed a year later as a joint venture of its own and Siemens'. The company inaugurated Fanuc Europe S.A. in 1978, the granting of licenses to Bulgaria in 1979 and to the People's Republic of China a year later, and the establishment of Fanuc Germany G.m.b.H. and Fanuc U.K., Ltd., in 1980 and of Fanuc Machinex in Bulgaria in 1981. In 1982 it entered a joint venture with General Motors to form GM Fanuc Robotics Corporation, and established Fanuc Mechatronics S.A. in Luxembourg. By now there are Fanuc France S.A.; Fanuc Oceania Proprietary, Ltd., in Australia and Fanuc Singapore. Fanuc's main shareholders are the original parent firm, Fujitsu, Ltd.; Fuji Electric Co., Ltd., and Siemens A.G.

new products just on the day shift. At night the entire factory works in darkness, unattended except for one operator monitoring the displays at a control center.* The Fuji plant has been in steady operation since 1980. It is a triumph of imagination, superb organization and confidence on the part of Dr. Inaba and his team to conceive and bring off such an achievement within the framework not of a subsidized project, but of a straightforward commercial investment supported by company profits.

From 1956, when Dr. Inaba's company shipped its first NC machine, until thirteen years later, Fanuc, Ltd., remained small, shipping just $10 million worth of products in 1969—comparable to a Silicon Valley start-up company. Then in the following six years the company's sales grew at an average 30 percent per year, to make it a $40-million company by 1975—still modest in size nineteen years after its first shipment of a product. But it achieved and held a 35-percent annual growth rate, and in six more years grew to sales of $373 million. And as sales grew, so did the company's before-tax profits: 17 percent in 1977, 25 percent by 1979 and a whopping 33 percent by 1981, when the effects of the new flexible manufacturing system at the Fuji plant began to be felt. Fanuc, Ltd., made that last factor of 6 in growth while paying average yearly taxes of about 50 percent. Even more significant was its growth in 1981, a time of worldwide economic stagnation. As Inaba said, "In the machine-tool field, both domestic and overseas demand continued to be active. . . . Demand for NC systems benefited from the fact that much of the new plant and equipment investment [by customers] was directed at saving labor."

As a result, in the stagnation period 1980–1981, Fanuc's orders increased in just one year by 45 percent, its sales by 64 percent and its profits by 71 percent. By 1981, it was producing a sales value of almost $400,000 per employee per year averaged over its entire payroll—a performance unmatched by any conventional "industry" except a few oil companies. But the company did not become simply a "cash cow." It continued to invest strongly in the development of new products, came out with a new System 9 series of computerized NC machines with simultaneous control of fifteen axes, and switched its regular line of commercial products from punched-

* I had some difficulty scheduling a time to visit the Fanuc, Ltd., plant by Mount Fuji because the company's managers were concerned about protocol: there would be no one senior enough to receive a visitor on the site during ten of its fifteen operating shifts.

paper-tape control, traditional in the NC industry, to a new system using magnetic-bubble memories.

The lessons of Fanuc, Ltd., should not be lost on us. First, there were the courage, the patience and the economic staying power that built a largely new technological base over a lean period of nearly two decades; then there was the close cohesion of management and labor that nurtured talent; and finally, there is the extraordinary payoff, in every sense, from being out in front of an industry after building for that position during a quarter-century. If America and Europe are to survive economically in the face of that kind of competition, they are going to have to find ways to maintain that economic staying power and team cohesion. The courage and the good ideas exist in abundance already.

EFFECTS OF AUTOMATED PRODUCTION

The most significant question of all in regard to robotics is the effect of automated production on people and on the production of wealth in a competitive environment. Some assembly jobs remain relatively difficult to automate. Toshio Numakura of Hitachi pointed out that wiring is hard to automate because it requires fast pattern recognition and a high degree of manual dexterity, combined with the ability to deal with a medium—the wire itself—that does not have a fixed, predictable shape. The difficulties of automating wiring tasks push the computer and electronics companies toward systems with larger printed circuit boards. There are more chips on each of those boards, but the chips can be inserted by robots, and there are fewer boards to be wired together by hand.

To assemble motors and compressors in a fully automated factory is much easier, and already the assembly of motors at Fanuc, Ltd., is done 100 percent by robots. Automation is high but not total in firms like Black & Decker and in automobile-engine plants. But even for difficult challenges like wiring, the trend is clear: to compete, one must find ways to increase the level of automation, either by more capable machines or by designing to avoid operations that cannot be done by machine. Numakura says:

> Robotized factories are one of the biggest growth markets that we see. You can't reject the introduction of robots, because they are so efficient that they give an unbeatable competitive advantage. Already the cost advantage for Japanese semiconductor chips comes

from our high level of automation, and eventually the bonding operation will be one hundred percent automated. Clothing will be made entirely by machines. As for the displacement of labor, my opinion is that it will be solved by the higher wealth produced in automated factories. It's a wealth-distribution problem.

Minoru Morita of Matsushita Electric expressed the same conviction. He pointed out that the strong Japanese drive toward development of the 64k and 256k RAM memory chips and of magnetic-bubble memories is motivated largely by the application of those devices to robotics. Morita also said:

The diversity of robots has to be in the form of software rather than machinery. We use forty different kinds of sequential robots now. In ten years we expect to reduce that to only ten different kinds. And we will be using ten to twenty times as many robots overall as we do now. To compete, our productivity per worker must rise by two to three times in the next decade. The barriers to an even faster robotization are in programming, in the changeover of factory layouts and environment, and in the time needed for retraining workers to become the supervisors of robots.

That is a very different attitude toward robotics than is found in some countries of the West. Among the Atlantic nations, many people think in terms of a zero-sum, static world in which the displacement of a worker from a particular job is evil in itself. A 1981 headline in the British magazine *New Scientist* said "WORKING WITH ROBOTS IS A BORE," and the subhead was "Robots can make life even less interesting for shop-floor workers. Unless unions have more control over decisions in factories, their members will not benefit from new technologies." The assumptions are that the total work force employed in factories should remain constant, and that unions are the only entities that can properly uphold the workers' interests. That resistance to automation, common in the United Kingdom, where an adversarial worker/management relationship has been endemic for generations, is more than rhetorical: in 1981, Britain had fewer than 400 robots in all its factories combined, while Japan had already passed the 12,000 mark two years earlier (using the same definition of "robot").

Writing in *The New York Times*, Harley Shaiken warned of a deterioration of the quality of working life that would result from the introduction of robots on any production line in which human workers were forced to maintain a constant pace set by the speed of the machines. But there are two alternatives to that: don't automate at all—in which case the entire work force becomes unemployed through the failure of its company to compete with automated factories abroad; or automate to such a level that no human worker remains directly involved in monotonous drudgery. The second is the Fanuc, Ltd., approach.

Shaiken made two suggestions that seem unrealistic: that "desirable" machine-loading and light assembly tasks be reserved for humans (although they are among the jobs most easily automated); and that productivity increases brought by automation be translated not into more competitive, lower-cost products but into a shorter workweek for the same pay. Both suggestions fail to recognize the already stagnant U.S. productivity growth rate, and that U.S. labor costs are about 70 percent higher than those in Japan. Writing in *The Nation*, Shaiken suggested that government should intervene to maintain the same levels of employment in each geographical area. In an article titled "Detroit Downsizes U.S. Jobs," he proposed that "Government [should] enter into the production of key automotive components such as engines. In such a public–auto worker partnership—a Federal Engine Corporation—jobs could be created and located in unemployment areas rather than exported." The difficulty is that every artificial barrier raised, legislative or otherwise, to the rapid introduction of automation puts industry at a competitive disadvantage.

The Japanese world-view assumes change, and further assumes both the possibility and the desirability of producing greater wealth. It also assumes, to a far greater degree than in the West, a mutual obligation between the labor force and the management of each company to increase productivity and never to lay off a worker. Michiyuki Uenohara of NEC said, "There is a common recognition that higher investment in new technology and new production facilities means higher productivity, higher value-added and, ultimately, higher rewards to all members of the enterprise." The article "Japanese Technology Today" said, "Lifelong employment makes perpetual and intensive investment in human resources possible."

Clearly, there is no way that any company can simultaneously robotize its assembly lines and maintain a constant work force without vigorously increasing its total production, and that means more wealth produced. Hiroshi Watanabe of Hitachi, Ltd., said:

As a manufacturing concern, we have never experienced trouble about the introduction of robots in our factories. They have always been welcomed. In view of the fact that factory jobs consist of re-petitive work which ignores human nature, use of robots that take over such work is a humane consideration. I don't see why this is opposed in Western countries.

That acceptance of robotics in Japan has produced an explosive growth rate in the robot industry itself: in 1980, called by the Japan Industrial Robot Association "Robot Diffusion Year One," the pro-duction of robots rose by 80 percent over the year before, and about 98 percent of it was absorbed by Japanese industries. A total of 19,900 units were added (in the Japanese definition of "robot") to the domestic installed capacity. Japanese workers, guaranteed long-term employment if their employers prosper, tend to welcome the introduction of more robots and look for ways to make the robots more effective. Wahei Saito of Kawasaki Heavy Industries, Japan's largest manufacturer of robots, says that the high rate of introduc-tion of robots into Japanese industry is a direct result of "the differ-ence in labor–management practices" between Japan and the United States. Jack Warren, CEO of Omark Industries, Inc., in Portland, Oregon, said:

There's not one of the machine-tool makers in [the United States] that couldn't build the automated factories I saw in Japan if they were willing to spend the money. But what really is more important is the relation of the Japanese management to the work force, a more effective approach to inventory management and the entire Japanese attitude toward productivity improvement.

Although the Japanese Government does not provide subsidies to manufacturers of robots, it does support the purchase of robots by Japanese industry. In "Robot Diffusion Year One," the Japanese Government provided tax advantages that allowed companies to deduct as much as 53 percent of the cost of buying a computer-controlled robot in the first year of its use. It also sponsored a number of programs to provide low-cost loans to small and medium-sized companies that wanted to install robots. As a result of this united effort and an attitude toward increased productivity that is shared by government, management and workers, Japan is scheduled to have about 32,000 robots installed by 1985, versus an optimistic maximum of only 8,000 in the United States by that year.*

Each nation is faced with a choice that rationally can be made in only one way. The irrational but tempting alternative is to impose many restrictions on the introduction of robots. All the restrictions seem praiseworthy at first glance—and all of them are equally disastrous in the long run. The fundamental necessity for a society concerned about the welfare of its citizens is to produce wealth, and that means seizing every possible advantage in a fiercely competitive world. Any delay, any restriction in the introduction of more efficient production methods decreases a nation's share of markets, forces the closure of its industries and destroys its opportunities to produce wealth for its citizens.

Americans are beginning to accept these realities, but too slowly. Successful competition for market share demands the nearly total replacement of humans by robots in the handling of materials and the production of finished products. The prospect of that replacement is frightening to many people, but we have carried it out successfully once before. We have automated agriculture to such an extent that great agricultural wealth is produced by only 2 percent of our population. The corresponding automation of manufacturing is already so far along that it has changed the demographics of our country. Since 1960, the fraction of our work force whose jobs are directly related to the assembly lines has dropped from 35 percent to 20 percent. The transition that remains, to a productive industrial society with only 5 percent or less of our work force engaged directly in production, is essential if we are to compete successfully with the Japanese.

Just as the agricultural revolution of the past century moved our population from the farmlands to the cities and suburbs, the new industrial revolution will uproot many families. But if the wealth earned by industrial success is not to be squandered on ghost towns

---

* While we think of robotics as a non–Iron Curtain phenomenon, the Soviet Union has automated its industries to a degree of which few Westerners are aware. The first Russian robot was not built until 1971, but by 1981 the total number of robots installed in factories in the U.S.S.R. was second only to the number in Japan, and substantially more than the robot work force of the United States.

surviving on a government dole, we must accept the need for some geographical displacement. Fortunately, this new revolution does not demand as much displacement of population as did the old. States and municipalities can choose by their own actions whether they lose or gain population. Areas that support and encourage new, highly automated industries will create new high-skill jobs and will grow. Areas that resist the introduction of robotics will stagnate.

To introduce the new breed of robots, and to do so without imposing great suffering on workers and their families, we must also carry out two other major changes. Both are much more likely to be accomplished by individual companies and individual towns than by national programs.

The first is to reforge the bonds of trust, mutual support and cooperation among labor, management and government that have been eroded by decades of confrontation law and confrontation politics. There can be no cooperation in an environment in which a worker fears permanent unemployment any time productivity rises. Somehow we must "close the loop" in the production of wealth so that when productivity increases are achieved, they benefit rather than threaten the workers in the company that achieves them. That is not an idealistic pipe dream. Companies in high-growth industries are able to create more jobs than are lost by the introduction of robotics, and they do so by maintaining or increasing their market share.

In more traditional industries, with a total market that is growing slowly if at all, successful competition in worldwide markets inevitably means some reduction in the total work force. This necessitates the second major change. To give workers the higher skills that are essential to their continued employability, we must provide them with high-quality education.

In American society, so much more mobile than the Japanese, it is less practical to provide adult education in the companies themselves. Here, when a worker has acquired valuable new skills, he may job-hop to a competitor.

There is a uniquely American way to solve that problem, just as the land-grant colleges were a uniquely successful way to bring about the automation of agriculture. It is our two-year community-college system. For example, in Phoenix, Arizona, there is a strong educational system called the Maricopa County Community Colleges. Local high-technology firms like Motorola, Control Data and the Garrett Turbine Engine Company support that system, and top executives from those firms contribute their time and expertise to improving it. For many American workers who need retraining as the new breed of robots takes over their dull, repetitive assembly-line jobs, community colleges like Maricopa's are providing the first real opportunity in a lifetime for a genuine education worthy of the name.

CHAPTER 4

GENETIC HARDWARE

On the wall of a Fifth Dynasty Egyptian tomb, from 4,400 years ago, a bas-relief illustrates the steps involved in baking bread and brewing beer—proof that humans were using microorganisms to produce food and drink thousands of years before microscopic life was recognized as such. Brewing and the leavening of bread are examples of fermentation—the chemical transformation of organic compounds through the action of enzymes (complex organic molecules) produced by living microorganisms. The yeast colonies that multiplied from cells in the air, or on the husks of plants, made the alcohol for the ancient Egyptians' beer and the carbon dioxide to leaven their bread. Fermentation technology was already 4,000 years old by the Fifth Dynasty. The bacterial conversion of wine to the dilute solution of acetic acid that we call vinegar was routine in Babylon 7,000 years ago; the bacterial production of lactic acid to preserve milk as cheese has an equally long history, going back to Neolithic times. Homely foods such as yogurt, pickles and sauerkraut are also products of fermentation, with histories only a little less ancient.

The foods made through fermentation are staples of our life, and the glamour products of microorganisms are the antibiotics. Generally, each one is made by only one microbial species, and there are many of them: 5,000 at present, with about 300 new ones being found each year. Over half of all these antibiotics are made by just one microorganism, the bacterium *Streptomyces*. Since 1928, when the first antibiotic, penicillin, was discovered by the Scottish medical researcher Dr. Alexander Fleming, the production of antibiotics has been improved, slowly and with much effort, by the classical Mendelian technique of screening. In biology, screening means testing many cell cultures in the hope of finding an accidental mutant cell with more desirable properties than its ancestors. By screening, over more than half a century, the production rate of penicillin per liter of culture broth has been improved 10,000-fold —simply because today's industrial strains of the mold *Penicillium chrysogenum* are so far evolved genetically from their ancestors.

Until recently, industrial biologists could work only within the limits of classical concepts of genetics known more than a century ago to Darwin and Mendel: that biological organisms breed true (fruit flies breed more fruit flies only, and elephants more elephants); that there are occasional variants (mutations) among the natural progeny of any species and that there is natural selection: some of the mutations give the organism a better chance of survival, while others doom that variant to extinction. But now industrial biologists can alter the processes of variation and replication deliberately rather than wait for chance mutations. The biologist and writer J. B. S. Haldane foresaw as early as the 1920s that what we now know as "genetic engineering" of microorganisms would greatly affect our lives. In 1937, in his book *Daedalus: or, Science and the Future*, he wrote that "the new brand of biochemistry which will, I believe, arise from genetics will be concerned largely with the stages of synthesis of macromolecules such as proteins. And its final goal will be the explanation and control of the synthesis of life."

If microorganisms replicated only by asexual division, as had long been thought, the production of new mutant varieties would be painfully slow. But early in the explosive development of microbiology that began after World War II, Joshua Lederberg discovered that some varieties of bacteria are sexual. That discovery opened the way to deliberate sexual matings between bacteria of different types, though still within the same species. (A species, in biology, is a set of organisms that can mate with each other.) With such "crossings," radically new varieties could be produced far more rapidly than by asexual division.

It had been learned by 1944 that heredity was carried by DNA (deoxyribonucleic acid). Then in 1953, J. D. Watson and F. H. C. Crick made a breakthrough discovery: the genetic heritage was coded linearly in a pair of strands of DNA twisted into the now-famous double helix, each strand holding a chain of genes like beads on a string. In humans, every cell carries a double helix whose strands are 100,000 genes long. For industrial microbiology, the key words are "cloning"—the splicing of a sequence of DNA from one kind of cell into the DNA strand of an entirely different host cell—and "expression"—mass-producing copies of a specific protein in the host cell by the DNA sequence that has been spliced into the host's DNA strand. And usually beyond the needs of industrial microbiology, there is "function"—the carrying out by an entire cell of an action that was made possible only by the foreign DNA sequence.

The rate of accomplishment in genetic engineering is awesome. In 1976, there was the expression of a protein by a DNA sequence taken from a prokaryote (in the particular case, a yeast cell) and spliced into the DNA of a eukaryote (a bacterium). It had been assumed until then that it would be difficult to transfer DNA between such very different life forms—one having its DNA in rings, and with no nuclei in its cells; the other with its DNA in chromosomes and a nucleus in every cell. In the same year, a mammalian gene from a rabbit was cloned. A year later, a human gene was cloned; and in 1978, function was achieved for a mammalian gene (that of a mouse) spliced into the DNA of a bacterium. In 1979, two groups of researchers announced simultaneously the expression of human growth hormone in a bacterium. Peter Farley of the Cetus Corporation said, "This is the Golden Age of biology."

When a cell divides, the DNA carrying its genetic inheritance replicates, making a "copy of the blueprint" to pass on to the daughter cell. The generation time is short: for the harmless intestinal bacterium *Escherichia coli*, long favored by researchers for its vigor and comparative simplicity, only 20 to 45 minutes is needed for copying the DNA blueprint and splitting to give two *E. coli* cells, each with a blueprint for further replication. Twenty-four hours of such replication, without losses, could yield 4 billion cells from the original 1. For industry, the lure of genetic engineering is the ability of the DNA not only to replicate itself but also to direct the production within the cell of complex proteins essential for life: insulin, antibodies to fight disease, and enzymes that catalyze chemical reactions.

In its mechanical structure, DNA resembles a long ladder gently twisted. The sides are made of sugars and phosphates. The rungs are formed by pairs of just four chemical bases (nucleotides): adenine (A), guanine (G), cytosine (C) and thymine (T). The base A pairs only with a T, and C only with a G, so there are only two different kinds of rungs. In DNA replication, a single strand carrying the sequence of half-rungs -A-T-T-G- picks up from the broth of unattached bases in the cell the same complementary sequence -T-A-A-C- that formed the other halves of the original rungs, while the second strand, carrying the original -T-A-A-C-, picks up bases to form the sequence -A-T-T-G-. In that way each strand builds a duplicate of its complement, and two complete ladders are made from the original one.

For many years the burning question in genetics was how the cell made proteins, which are complicated molecules built out of just 20 amino acids. Typically, from 100 to 300 amino acid molecules make up a protein. Taking the four bases as letters, a sequence of any three forms a "codon" which directs the cell to make a specific amino acid. As there are four different bases in the rungs of the DNA ladder, the number of different codons, each a sequence of three bases, is $4 \times 4 \times 4$, which is 64—far more than the 20 amino acids. There is redundancy. Though certain amino acids, like tryptophan, are coded by just one sequence of bases, others, like cerine, can be coded by any one of six different sequences. Of the 64 possible sequences, three are reserved to mean "Stop!"—stop making the protein. As in a factory, where the master blueprint is not exposed to dirt and possible damage, the DNA ladder does not make the amino acids directly, but rather makes working drawings, "messenger RNAs," each of which has a strand carrying the useful information and a "nonsense strand" that is just a template. Typically, 300 rungs on the DNA ladder form a "sentence" of 100 words that make up a gene.

The particular sequence of amino acids in a protein determines its three-dimensional shape, and some proteins are in our bodies simply to provide structure: collagen for muscle, keratin for skin. Some are hormones, like insulin. Others are antibodies, which work by grabbing a virus or bacterium and rendering it harmless until a white blood cell absorbs them both. But the vast majority of proteins are enzymes, organic catalysts that enable and control the thousands of reactions within a living cell. Many of the enzymes control energy exchanges. Sugar is the "gasoline" of the cell, but the cell could no more survive the direct burning of sugar than a car could survive having a match tossed into its gas tank. Instead, the cell carries out the breakdown of sugar into carbon dioxide and water by separating the process into dozens of intermediate reactions, each with a small, tolerable release of energy. For each of those reactions, there must be a specific enzyme ready at the right time and place. That, in short, is life.

RECOMBINANT-DNA TECHNOLOGY

The primary techniques for recombinant-DNA technology were developed in the last fifteen years. Hamilton Smith of Johns Hopkins University discovered the DNA code sequence of a "restriction enzyme" in 1970. The enzyme cut the DNA ladder at a specific codon. Later he found an enzyme that cut the ladder at a point leaving two "sticky ends" to which foreign DNA could be attached. Daniel Nathans, also of Johns Hopkins, used this enzyme to make virus recombinants. By 1972, Paul Berg of the Stanford Medical School had found another, independent way of splicing genes. In his method, with the help of "end-addition" enzymes the DNA ladder extended itself controllably. All three men were awarded Nobel Prizes.

Later, Dr. Stanley Cohen of Stanford identified in *E. coli* a small free-floating ring of DNA called a "plasmid" that was not a part of the bacterium's chromosome and would be ideal for carrying a spliced-in gene. Dr. Herbert Boyer of the University of California had found another restriction enzyme, Eco R1, which cut the DNA ladder at a specific codon, leaving two "sticky ends."* Eco R1's talent lies in its cleaving the plasmid at an axis of symmetry, where the sequence of bases reading leftward on the top strand is the same as the sequence reading rightward on the lower strand. Here is an example. Before cleavage, the strands look like:

-G-A-A-T-T-C-
-C-T-T-A-A-G

Cleavage by Eco R1 slices the DNA ladder's side bar between a G and an A, then cuts four rungs, and completes the cut by severing an A and a G, so that the left and right open ends are left as:

* By now more than 400 restriction enzymes are known, of which about 80 are available commercially.

110

$$-G \qquad\qquad and \qquad A\text{-}A\text{-}T\text{-}T\text{-}C\text{-}$$
$$-C\text{-}T\text{-}T\text{-}A\text{-}A \qquad\qquad\qquad\qquad G\text{-}$$

Because there is no difference between "top" and "bottom" as far as the sticky ends are concerned, the two ends are identical in their subsequent attachment. As an example in terms of language, the cleavage has been made at the center, or symmetry line, of a "palindrome"—a sentence that reads the same either left to right or right to left. The analog of the cleavage line is "R" in Napoleon's "ABLE WAS I ERE I SAW ELBA." With Eco R1, an identical cleavage is made in the gene whose replication is desired—for example, the gene for making insulin—and the cleaved plasmids and genes are mixed. Occasionally a plasmid and gene join at their sticky ends to form a complete ring, capable of reproducing itself.

At first the recombined DNA is frail, with the severed rungs cemented only by the weak bonding of the nucleotides' hydrogen atoms. But then the genetic engineers introduce another enzyme, ligase, which "anneals" the bond by adding the sugar and phosphate backbones to complete the ladder. The recombined plasmid breeds true, reproducing with each 45-minute generation of its host *E. coli*, so that in a few hours the engineers have thousands of the cloned bacteria, each with a gene producing the specific protein they have chosen.

THE CETUS CORPORATION

To understand the explosive growth of genetic engineering, we will follow the history of one genetics firm. The professional career of Ronald Cape, Chairman of the Board of the Cetus Corporation, began in the 1950s, just as the first discoveries in biology's explosive growth were being made. Cape, energetic, compact and ebullient, was born in Canada and graduated summa cum laude in chemistry from Princeton University in 1953. Later, he took an M.B.A. at the Harvard Business School. When Sputnik went up in 1957, he became restive at the prospect of a new world of applied science opening without him. Four years later, in the U.S. science exhibit at the Seattle World's Fair, Cape's itch became a fever when he saw at a demonstration for the public the awesome experiment of Meselson and Stahl on the separation of the DNA strands and formation of complements of themselves—key to the replication of DNA.

As Cape said, "That demonstration absolutely propelled me back into school."

By 1967 he had earned a Ph.D. in biochemistry from McGill, and he followed it by studying the genetics of phages—the viruses that infect bacteria—at Berkeley. By then it was 1970, and Cape's unorthodox blend of science and business education, looked at askance fifteen years earlier, had turned into a plus.

With Dr. Peter J. Farley, who held both medical and business degrees, he formed the consulting firm of Cape-Farley. The two partners set out to advise prospective investors on the new field of molecular biology, and began by reviewing more than a hundred applications from entrepreneurs hoping to start new firms. They turned down every one, and found in their review that generic corporate names were easily forgotten. Later, when the time came for them to set up a firm of their own, they chose a name that was quirky, memorable and totally unrelated to microbiology. Now, more than a decade later, Cape muses on the scientific advances on which his business career is based, and he still marvels:

> It's fascinating to discover how mechanistic biology is, at its most fundamental level. The construction of a cell, the way information is stored—they're exactly as an engineer would have designed them. Nature uses blueprints, working drawings, switches, codes and feedback. Those aren't metaphors—they're physical realities. One by one, the predictions that Francis Crick and his contemporaries made years ago are being borne out: the triplet nature of codons, the existence of the transducer step that we call messenger RNA, the existence of restriction enzymes and repressors.

The powerful and precise repressors, now used as cutting and welding tools by genetic engineers, are thought to have been developed by certain bacteria as insidious weapons to penetrate host cells and turn the host's DNA to their own advantage. There are 61 different messenger RNAs, one for each of the 61 codon triplets. Each of the messenger RNAs is designed with a combination of geometrical shape and a distribution of electric charges to fit and hook onto a particular amino acid, with the same 61-to-20 redundancy that the codons exhibit.

"There's even multiplexing," says Cape—

> like the way a communications engineer crams extra signals into a limited band width. Some viruses have very little DNA, and they

112

multiplex by reading the same DNA strand three different ways, getting sense and coding with each.

He means reading a given strand from three starting points, each displaced by one nucleotide from the last. If we imagine the nucleotide "letters" in their numerical sequence this way on the strand:

1 2 3 4 5 6 7 8 9 10 11 12

then the virus reads one set of codons by associating the nucleotides into the sequential triplets

1 2 3, 4 5 6, 7 8 9, . . .

But it gets another set of triplet codons by starting one place to the right, as

2 3 4, 5 6 7, 8 9 10, . . .

and it gets still another set of codons by reading the sequence as

3 4 5, 6 7 8, 9 10 11, . . .

so that it can encode a "program" as complex as if its DNA strand were three times as long. Continuing the list of mechanistic components in the cell, Cape likens the ribosome "jig" to the moving slide of a zipper. It couples to one after another of the codons along the line of the messenger RNA strand, does its job and leaves a copied codon. It holds up to three amino acids in its jig at once, and continues along the messenger RNA strand until it hits a "stop" sequence.

"And the other surprise," Cape continues,

is that the great barriers we imagined were largely in our minds. We can cross species boundaries now. We can take the sequence from the human chromosome that codes for insulin production, clip it into the DNA strand of a bacterium and have that bacterium produce human insulin. And the "barrier" between prokaryotic and eukaryotic forms of life—it doesn't exist, at least as far as insulin production is concerned. Now we can do "reverse genetics." We can use all we know about genetics and biochemistry to predict the organism we need, and then change the DNA structure to create that new organism and make it breed true.

Cape turned from consulting to manufacturing in 1971, by forming the Cetus Corporation in partnership with Peter Farley and Donald Glaser. Glaser, who won a Nobel Prize in physics for his invention of the bubble chamber, a particle-tracking device, is a professor at Berkeley in both physics and molecular biology. His specialty is the screening of microorganisms to find rare, useful mutated strains. Cetus, the oldest and the largest of the independent genetic-engineering firms, made its first profits from the licensing of technology for large-scale, highly automated biological screening. The Schering-Plough company pays royalties to Cetus on a microorganism that Cetus developed with the help of those techniques, an organism that Schering uses for the production of the antibiotic Netromycin.

While Cape, Farley and Glaser were choosing a name for their new corporation, a movie, *The Andromeda Strain*, was frightening audiences with a fictional tale of a crystalline, alien life form run wild. Glancing at a star chart, the partners noticed the constellation Cetus (the Whale) next to Andromeda. Cetus evokes pleasant connotations, and the name is nongeneric—it has nothing at all to do with genetic engineering.*

Cetus' advertisements define bio-industry as "the innovative use of microorganisms in the profitable production of materials to fill human needs." The emphasis is on the "profitable." Says Cape:

We make the same choices as a chemical company. Usually it's business reasons, not science, that rule our decisions. Like how much capital it will take to bring a product to market. Or how much competition there's going to be. If you try to break into a market

---

* In Cetus' Berkeley, California, headquarters a certain whimsical, self-mocking humor is carried a step further. Entering, the visitor notices an oil painting that covers a whole wall. In a style combining realism and fantasy, it shows a molecular biologist, easily recognized as Francis Crick, in his laboratory just as a tidal wave breaks through the walls, carrying pelicans, herons and a friendly whale. The scene has, as Cape puts it, "a certain irreverent charm."

where somebody else has an eighty-percent share, they may drop the price just long enough to bankrupt you.

The specialty publication *Chemical Week* called Cetus "uncannily perceptive" in company financing, and Cetus has become the largest firm devoted exclusively to biotechnology. When Cetus was formed, its founders bankrolled the first year of operations themselves. Then they sold 20 percent of the ownership of the company for $2 million to professional investors. A year later they brought in another $3 million, but for only a further 8 percent of the ownership. To spread risks, obtain working capital and tap into the experience and expertise of corporate giants in the chemical industry, nearly 28 percent of the company was sold to Standard Oil of Indiana in 1977 for $5 million. The oil company asked for and secured two places on the board of directors in the deal. A year later, on the same terms, Standard Oil of California bought 22.4 percent of Cetus. Soon afterward, National Distillers and Chemical bought 14.2 percent of the Cetus stock, and was assigned one seat.

The apparent control of Cetus' board of directors by giant firms holding five of nine seats evidently didn't worry Cape and his fellow scientists on the board, but they were concerned about the need to satisfy the firm's voracious appetite for research-and-development funding through a period of years when earnings would be meager. Their decision, brilliantly timed, was to go for all they could get in a public offering. According to Cape, that move, made in 1981, was "the largest initial public offering in history. And it hasn't been topped since" (as of 1982). In the new offering more stock was issued. It was sold for $107.2 million, reduced the holdings of the industrial giants to just under 50 percent of the total company, left Cetus with no long-term debt and enabled the firm to sit back for another four to five years without refinancing. In *Chemical Week*'s analysis, that move put Cetus in a stronger position than its nearest rival, Genentech, which had been able to raise only $36 million in a similar offering. Genentech, a pioneering firm, had been the first genetic-engineering company to go public in a big way.

About one quarter of Cetus' financing is being used to fund its own research, so that the company is not being forced to bring products to market with excessive haste. Another 10 percent went into the financing of two subsidiaries, each led by a distinguished academic scientist. The Cetus Immune Corporation in Palo Alto, directed by Dr. Hugh McDevitt, in 1980 began selling through a larger company, Cooper Laboratories, an antibody that is useful in the diagnosing of certain diseases before organ transplants. While sales will always be small, the antibody is significant as the first product of genetic engineering to reach the marketplace.

Cetus Palo Alto, directed by Dr. Stanley Cohen, is developing a battery of biologically active tests that can diagnose toxoplasmosis, a disease which, if contracted by a mother during the first trimester of pregnancy, could damage the fetus. Still another subsidiary, Cetus Madison, was formed in 1982 to exploit the opportunities for genetic engineering in products for the agricultural industry.

But Cetus remains a relatively small company, and Cape sees its future largely in terms of joint ventures with larger firms. The prospective big-dollar products of those ventures are, for the most part, simple industrial chemicals rather than wonder drugs. Since 1977, the French company Roussel Uclaf has used a Cetus-developed process for the production of the vitamin B_{12}. Patents have been filed on behalf of the Standard Oil Company of Indiana for a Cetus-developed biological process that yields xantham gum, a binder that increases the recovery of crude oil from wells. And National Distillers is building a plant for $100 million—comparable to the entire book value of Cetus—to produce fuel alcohol by a process that Cetus developed.

One of the biggest challenges that Cape and his colleagues are now targeting is even more pedestrian—the production by microorganisms, at lower cost and with less pollution, of the oxides and glycols of the basic petrochemical feedstocks ethylene and propylene. Those chemicals are the input materials for automobile antifreeze, polyester plastics and urethane foam—products whose world market exceeds $5 billion annually. In a Cetus process described by *Business Week* as "elegant in its simplicity," three bacterial enzymes, coated on glass beads in a column through which ethylene or propylene gas is pumped at room temperature and low pressure, act as catalysts to enable and direct the reactions desired. The new method replaces a traditional technique that required high temperatures and pressures and expensive metallic catalysts. Peter Farley of Cetus calls that venture "the first substantive case where microbiology will take a shot at the heart of the chemical industry."

Although the potential is indeed great for such new processes, their introduction will be slow because the chemical industry has more capacity for present production than it needs, in the form of old plants whose costs have long since been written off. In spite of that barrier to innovation, more than two hundred genetic-engineering firms have been formed, all aiming at the pharmaceutical or chemical industries. A few of them—Cetus; Genentech; Genex; Biogen S.A. of Geneva, Switzerland, and several smaller rivals—have the resources to wait out a long dry period during most of the 1980s, when development investments will be high and profits relatively small.

AN "ANDROMEDA STRAIN"

The prospects for companies like Cetus would not be so bright if concerns about recombinant-DNA research were as strong now as they were in the early 1970s. Then many people, among them the recombinant-DNA researchers themselves, feared that a bacterium genetically engineered to be resistant to disease would escape from the laboratory and infect the population with a plague for which there would be no cure. As a group, the community of DNA researchers voluntarily refrained from such "dangerous" lines of research until new "anti-Andromeda" laboratories could be built.

One such facility resembles a movie set for a futuristic submarine: multiple doors with thick seals are set into massive hollow walls filled with a liquid plastic that automatically expands to seal any puncture. Responding to the strong concerns of the early 1970s, the National Institutes of Health (NIH) published a set of strict guidelines for the conduct of DNA research. Towns and cities such as Boston and Cambridge, Massachusetts; Berkeley, California, and Princeton, New Jersey—all with universities and all the centers for research-oriented companies—passed local ordinances that turned the guidelines into laws. But as recombinant DNA became less mysterious, the fears subsided. No real-life Andromeda Strain has emerged, and some geneticists argue that microorganisms have been using recombinant-DNA techniques for millions of years. The NIH guidelines have been relaxed, and the Cetus anti-Andromeda facility, built at a cost of $2 million, was turned over to the research staff to be used most of the time as just another clean room.* About 80 percent of all recombinant-DNA research can now be done with

* The NIH guidelines are mandatory in federally sponsored research, but voluntary for industry.

the ordinary sterile procedures that are routine in hospitals. But if any company wanted to pursue research on the DNA of disease-causing bacteria or viruses, or on genes for the synthesis of strong poisons, that research would still have to be carried on in top-security "P4" labs. As of the early 1980s, no such research was being carried out by industry.

RESEARCHERS AND PROFIT

While fears of a genetically engineered plague organism's escaping the laboratory and decimating the population may have been exaggerated, less dramatic negative aspects of genetic engineering are now becoming apparent. Most are the result of recombinant DNA's power and success. Because it works, and has great potential for profit, it has become an irresistible lure for academic researchers who had grown accustomed to, but never really satisfied with, the meager financial rewards of their traditional academic jobs. Many of those researchers left their ivory towers to form or join new companies. Walter Gilbert, for example, a mathematician turned physicist turned molecular biologist, shared the 1980 Nobel Prize for chemistry, and then a year later left Harvard to become Chief Executive Officer of the Swiss firm Biogen, and to supervise the building of a new Biogen laboratory in Cambridge, Massachusetts.

The brain drain of scientists from academia to industry is not easy to stop because, as the British scientist Cesar Milstein observed, "In this society you're made to feel stupid if you can't make money." During the 1980s, universities will be wrestling with the choice between taking a hard line—forcing their entrepreneurial professors to sever all academic ties—and keeping them on some loose, nontraditional rein in order to benefit from their expertise. Two of the best universities that are centers for genetic-engineering research, the University of California and Stanford, have concluded that they gain more than they lose by permitting their professors to take on responsible roles in nearby industries. Switching fields is hardly new. Many creative scientists change fields entirely several times in the course of their careers. But given the high salaries to be earned in recombinant-DNA companies, those who join them will find it difficult to make the sacrifices attendant on a return to academia.

The aura of wealth that now surrounds genetic engineering is having an unfortunate effect on governmental funding in biology.

As Ron Cape notes, corporate decision making will always favor the relatively short term. Truly basic research, with no potential for payoff within ten years or more, will remain unfunded unless the government picks up the tab. It takes an observer far more learned than the typical government official to look beyond today's boom in genetic engineering and appreciate that no one is feeding the goose of pure research that eventually lays the golden eggs.

THE IMPACT OF GENETIC ENGINEERING

Genetic engineering is certainly going to make an impact on our lives in the next ten to fifteen years because of its production of drugs for the treatment of disease. But there is a curious paradox. To a small number of people who could only have looked toward an early death, it will give life. That effect is immeasurable in human terms, yet will be relatively small in economic terms. When we search for those products of genetic engineering on which to base whole new industries of major scale—at least tens of billions of dollars per year—we find no promising candidates likely to become important before the turn of the century. Cape has a small, 10-liter reaction vessel sitting on a laboratory bench, with tubes and sensors connected to it and a computer nearby to control its valves and heaters. He said:

> This is our interferon pilot unit. In this one vessel we can make ten thousand patient-doses of interferon overnight. We take a human gene, stick it into a bacterium and say "Do it!" And the project is real human interferon, not a close animal relative of it. Interferon is a hot prospect for the treatment of cancer, and possibly for other "autoimmune" diseases—diseases in which your body's own chemicals are destroying you: muscular dystrophy, maybe multiple sclerosis, probably rheumatoid arthritis . . . Tay-Sachs disease and sickle-cell anemia. Many and maybe all virus diseases can be attacked by natural products of the human body. Almost everything that kills people except heart attacks and car accidents potentially can be attacked by these methods. People are going to live longer.

And—of potentially greater importance—people are going to enjoy better health in the last decades of life.

DRUGS AND THEIR MARKETS

But for all Cape's enthusiasm, he does not see genetic engineering yielding any really big industries until well into the 1990s. Some diseases, whatever the anguish they cause for individuals, affect so few people that drugs to cure them will never constitute a large market. Cetus Palo Alto's test for toxoplasmosis will fill a market estimated at less than $20 million worldwide, and most other sophisticated products of genetic engineering aim at markets of comparable size. Even if interferons turn out to be effective on the widest possible scale in treating cancer and other diseases, the very success of genetic engineering in giving high yields at low cost will force their price down in a competitive market.

Also, no one will get rich quickly by manufacturing new drugs, because the governmental regulatory process to ensure safety (a process that Cape and his fellow manufacturers endorse) forces a waiting time of nearly a decade for each new human drug while tests on it are carried out. For drugs used on animals the wait is not so long, but there are no huge markets. For example, using recombinant-DNA technology Cetus produces a vaccine that prevents scours, a disease that, untreated, kills 10 percent of all newborn swine and cattle.* Saving the lives of 10 percent of all piglets and calves born seems economically important, but in fact it is estimated to be worth only $60 million in drug sales worldwide.

Larger opportunities, but with correspondingly heavy capitalization, exist for more mundane products. Vitamin C, sold over the counter, accounts for several hundred millions of dollars in sales per year, worldwide. Recombinant-DNA research may yield a timesaving process for producing it. All together, the market for drugstore products adds up to about $60 billion; but almost 90 percent of that market is made up of soaps, cosmetics and other over-the-counter products of low intrinsic value for which genetic engineering could do little. And of the small fraction of the $60 billion represented by drugs, much goes into production machinery

---

* Human and animal vaccines have always been tricky to use. If killed, they lose potency. If only attenuated, they can produce the disease they were designed to prevent. Such an episode crippled many children when vaccines for polio were just being developed. Recombinant-DNA products contain only a portion of the microorganism, so they cannot replicate in the body. But that portion is enough to stimulate the production of antibodies.

and a diffuse infrastructure of sales, not into the proprietary aspects of the drugs themselves.

For producing insulin, interferon, cattle vaccine and most other recombinant-DNA drugs, genetic-engineering companies plan to license their methods to established, heavily capitalized pharmaceutical and chemical companies. The royalties earned by the recombinant-DNA wizards are unlikely to be more than 10 percent of the corresponding world markets. But appropriately, to the few individuals who played key roles in the start-up phase of genetic engineering, hard work and vision have brought substantial wealth. The founders of Cetus each own about 4 percent of its stock. In 1975, when the rival firm Genentech was founded by Robert Swanson, then 28 years old, and Dr. Herb Boyer of the University of California at San Francisco, each of the two put up $500 in start-up capital. Five years later, when Genentech went public, Boyer's resulting 925,000 shares in Genentech were valued at $35 million. As a Kidder, Peabody analyst noted, "The main assets of the genetic-engineering firms in these developmental years are their scientists. The trouble is, those scientists have legs and they can walk."

Cetus, Genentech and most of the other DNA firms try to anchor their scientists with attractive stock-option plans. About 12 percent of Cetus is owned by its employees—an amount comparable to the founders' holdings. But the considerable wealth gained by individuals in the first years of genetic engineering obscures the fact that so far not much in the way of true wealth generation has occurred.

The long lead time between development and products in genetic engineering poses a difficult management challenge for the companies. The top four identified total markets for drugs that the genetic engineers can make are not very large compared with the markets for the products of microengineering. In dollars per year, those total markets are estimated to be: human interferon, 0.5 billion; human insulin, 0.3 billion; hepatitis vaccine, 0.2 billion; infectious-disease diagnostics, 0.2 billion. As a Cetus stock offering notes:

> The worldwide insulin market is dominated by two firms and success in it is related to marketing rather than production skill. We expect Cetus to license its insulin-making expertise to one of those current market leaders.

The U.S. market for prescription drugs is about $8 billion annually, and the long-term potential is for a 20-percent penetration of that market by the products of recombinant DNA. Earnings in the larger markets for commodity chemicals appear to be a long way off. Dr. Zsolt Harsanyi, writing for the congressional Office of Technology Assessment (OTA), said that the promise of genetic engineering is nothing less than "the possibility of building a sustainable future based on renewable resources." But as the genetic engineers recognize, making a big impact on the commodity-chemical industry is vastly different from turning out 10,000 patient-doses of interferon overnight in a 3-gallon jug. As Ron Cape says:

> The existing cover stories are about potentially enormous markets, like turning agricultural products into fuels. Nobody's going to make big dollars on them for fifteen to twenty years. Not because of scientific problems—the science is so damned powerful that it's rarely the limiting factor—but because of pedestrian economic problems. We can turn any biomass into liquid fuel on the lab bench, but that's the smallest part of the problem. There's no equivalent for biomass of the whole infrastructure of ships, pipelines and refineries that's been built up over sixty years for the petroleum industry. Building the infrastructure for biomass will take a full generation. That's why talk about "gasohol" for this century is nonsense—it's just pandering to the farmers.

The commodity-chemical markets in which the DNA firms can reasonably hope to compete are large by the standards of small start-up firms, but small on the scale of the major industrials. Propylene and ethylene oxides find uses not only in antifreeze and industrial plastics but as stabilizing agents in foods, cosmetics and tobacco; yet their annual U.S. market is estimated at only $1.5 billion. Some genetic engineers estimate that DNA manipulation could eventually be applied to about 25 percent of all chemical production, affecting $40 billion of annual product. The OTA assessment is significantly smaller: out of a total annual chemical market of $150 billion, the OTA estimates that only $11 billion can eventually be affected by DNA engineering. And unlike microengineering or robotics, where whole new multibillion-dollar industries spring up within a few years to manufacture products that did not exist previously, the recombinant-DNA contribution to the chemical industry during the next ten to fifteen years appears to lie in relatively marginal improvements in production cost for products that are already well established.

AGRICULTURAL APPLICATIONS

Dr. Ron Cape of Cetus is fully aware that DNA engineering for human and animal drugs is unlikely to become really big business before the 1990s. But he thinks that in agriculture the economic opportunities are substantially greater. Dr. Howard Schneiderman, Senior Vice-President for Research and Development of the Monsanto Chemical Company, takes an opposite view: that DNA engineering will not yield big new markets in agriculture for at least another decade. Schneiderman says:

> For health care, the total pharmaceutical market is $50 to $60 billion per year. The agricultural-chemical business, other than fertilizer, is just $10 billion. And the seed business is only $3 billion. I think those relative proportions are going to stay the same, even twenty years from now.

He sees big opportunities for genetic engineering in human health-care products, especially for the treatment of allergies, rheumatic diseases and certain genetic defects. Spliced-gene drug products would be engineered, he foresees, to control appetite, mental illness and extreme rage. However, his more cautious view of the opportunities in agriculture for DNA engineering does not mean that Monsanto is uninterested in them.

Monsanto is a major force in conventional agricultural chemicals, which account for more than $1 billion in its annual sales. Its herbicides are sold in more than 100 countries. Monsanto has a firm commitment to the new field of biotechnology. It employs about 200 people in that field, and is building a new $185-million life-sciences research "campus" 10 miles from downtown St. Louis, to support an eventual staff of 1,000 people in biotechnology. In 1982 the firm also announced a $23.5-million joint program for new-drug discovery, in cooperation with its neighbor Washington University.

Yet Monsanto's commitment to conventional agricultural research will be even larger. In Schneiderman's view, the nearest-term application of DNA research to agriculture will be in animal growth hormones. They will be on the market before 1990, he estimates, and will result in higher feed efficiencies, so that animals will yield more milk and more meat for the same amount of feed. In the plant world, he sees DNA engineering being used first to help understand how a plant works, and next not on the plant itself but on the bacteria that surround it.

A growing plant is really two creatures, the green stalk and the tiny partners that surround its roots—the microorganisms. They have many roles beyond nitrogen fixation. Because bacteria have relatively few genes, genetic engineering on them is fairly easy. I can see that happening by 1990.

But Schneiderman estimates that using recombinant-DNA technology to improve seeds will be more difficult, and will not happen until the last decade of this century. "So far it has been possible to grow an alfalfa plant that has been worked on by genetic engineering," says Schneiderman, but (as of 1982) "no one has been able to do that with one of the prime agricultural plants." In total crop value, alfalfa ranks number three in the United States, and is used almost exclusively in animal feed. For crops eaten by humans, there is the question whether people will choose to eat them. Corn, for example, is a good and palatable food, but it is low in the essential protein lycine. Years ago a new, high-lycine corn was developed, and every lab test showed that it was more nutritious than ordinary corn. Unfortunately, no one bought it, because it "looked funny." As for still more challenging tasks like resistance to disease, Schneiderman believes that even in the ideal case where disease resistance depends on just one gene, to transfer it from one plant to another— for example, from grass to wheat—will take about ten years.

Schneiderman agrees with Ron Cape that the fermentation of biomass to produce fuels and other commodity chemicals is

a next-century thing. Methanol production from coal is going to be very important, and fermentation won't be able to compete with it. You can't store cornstalks. And petrochemistry will continue to dominate, because it gives an unbroken array of molecules all the way from one to twelve carbon atoms long. Of course, there may be local exceptions, like Brazil, where a local surplus of labor may make it profitable to convert biomass into ethanol.

INTERNATIONAL COMPETITION

Dr. Schneiderman believes that in molecular biology and genetics, the United States is the world leader by far. But he notes that the European research in genetic engineering is very good. Imperial Chemical Industries in Great Britain has the largest single-tank fermentation plant in the world for the production of proteins. And Hoechst in Germany is the largest drug company in the world. The firm is particularly strong in value-added products, and does less in basic chemicals. Hoechst estimates that the world market for ethical pharmaceuticals in the year 2000 will be $200 billion.

Schneiderman and Cape agree that in fermentation engineering Japan is the world leader by far. The Japanese will be using genetic engineering particularly in trying to improve the yield and quality of rice. As to our competition with the Japanese, Schneiderman says:

> I'm very troubled about the overall question of how U.S. industry can compete with them. The absence of "stock analysts" over there makes it much easier for Japanese firms than for us to invest in the future. As for the work force, their literacy is way over ninety-nine percent, while ours is going down. And neither the U.S. Government nor U.S. business appreciates how tough the competition with Japan is going to be. It's one thing to compete with another company, but it's quite another thing to compete with a whole country. America is traditionally good at emergency solutions, but much poorer at long-term solutions. I'd hesitate to predict the outcome.

One can gauge the extent of European and Japanese commitment to biotechnology directly by the list of joint ventures. While some U.S. DNA firms have linked only to larger U.S. corporations, Genentech has acquired an international flavor. Its genetically engineered human growth hormone, for the treatment of hypopituitary dwarfism in children, is being brought to market in a joint venture with KabiGen, a Swedish pharmaceutical manufacturer. Genentech will produce human immune interferon in a joint venture with two Japanese companies, Daiichi Seiyaku and Toray Industries. Both are chemical firms of substantial size and considerable diversification. For example, in addition to its interest in pharmaceuticals, Toray Industries is probably the world's leading firm in the manufacture of ultrahigh-strength carbon-composite plastics. Another Japanese firm, the Mochida Pharmaceutical Company, Ltd., is also a leader in recombinant-DNA development.

In keeping with national policies favoring strong governmental roles in development, several European nations have invested government money directly in their genetic-engineering firms, or have set up nationalized laboratories. Celltech, a British firm, has a $28.5-million initial investment from the government's National Enterprise Board as well as from four large British investment firms. France has established a national development laboratory, the Groupe Génie Génétique, out of three large state-run research organizations and the Pasteur Institute.

The United States has a major competitive lead in genetic hardware as a result of funding for basic research by the NIH and the NSF. But that funding is now declining. Of the Europeans, Cape says:

> They talk "entrepreneurship," but it's the same old establishment —planned economies, government commissions to develop the "entrepreneurs." People who need that aren't going to be very tough competitors. The Japanese are another matter entirely.

He points out that the Japanese have a long head start in production know-how, because they have the biggest fraction of their GNP in fermentation industries of any major industrial power—7 percent. The products are unglamorous things like miso, soy sauce, food enzymes and beer—but the technology is just what will be needed: fermentation engineering. And the Japanese Government is encouraging those industries in every possible way. The firms also have abundant internal capital, because, just as in car production, they can get high leveraging and low interest rates.

"As for the leading edge in new scientific developments, that's a different story," says Cape.

> The Japanese say they intend to become innovative in that area, and I believe it's always best to view your competitor in the most respectful light. But for them to equal us in basic science, they're going to have to develop a society in which a twenty-one-year-old graduate student can tell his professor to "shove it!" because he's got a better idea. Maybe they can develop a subculture where that's possible, but it will take a long time. So for quite a while we'll have the bargaining chip of innovation to trade against their super ability in applications. I hope we use it.

126

Finally, Ron Cape believes that the only way American industry can be strengthened in its competition would be through lifting of the capital-gains tax entirely for money left invested in a firm through a full ten years—and that the law establishing that approach should be made to have a ten-year irrevocable minimum lifetime.

The future of genetic engineering as seen by Cape, Schneiderman and the OTA appears clear and consistent. To individual human beings suffering from diseases that once were incurable, gene splicing may give life. But for all the drama of the wonder drugs that recombinant-DNA research can produce, their overall economic impact will be much smaller than the earnings from microengineering and robotics. Eventually there may be new markets of more than $10 billion annually as a result of recombinant-DNA research; but those markets will probably be in areas like synthetic fuels or commodity chemicals, not in pharmaceuticals.

To develop those markets will require a very long time, probably several decades. And they will develop at all only if the price of fuels and other chemicals derived from wood or other plant products can be proved to be substantially lower than the price of the same chemicals derived from coal. The experts are doubtful that such a case can be made with enough certainty to justify the heavy expenditures of capital, maintained over a long period of time, that would be needed to build the infrastructure for a nation deriving its fuels or major commodity chemicals from biomass conversion.

In basic research, the United States clearly leads, but Europe is a strong contender. And in production engineering for the products of recombinant-DNA research, Japan is very strong, with Europe and the United States well behind. As the examples of Genentech and Cetus indicate, the glamorous new start-up firms that specialize in genetic engineering are likely to succeed mainly through joint ventures with long-established pharmaceutical and chemical companies, both here and abroad. In our search for potential really big new markets, we must look elsewhere than genetic engineering.

In the next three chapters, we will explore three opportunities that could spark the growth of new industries of very large scale, in the range of $10 billion to $100 billion or larger.

CHAPTER 5

MAGNETIC FLIGHT

On an island in Tokyo Bay, one sunny March morning in 1980, I joined other passengers aboard a streamlined wingless vehicle built by Japan Air Lines. The operator touched a small lever marked LIFT, and silently the vehicle rose by just half an inch, floating freely. The throttle was advanced, and the vehicle accelerated without any sound or vibration. On our "train" journey of several minutes, no wheel turned. There was only one moving part—the vehicle itself.

Two years later, I climbed to an observation platform 30 feet above the shoreline on the southern Japanese island of Kyushu, near the holiday town of Miyazaki. Far to my left, an elevated concrete guideway began among palm trees, at the research center of the Japanese National Railways (JNR). In a few moments, a full-size two-car railway train grew from a dot to a streak, and went by only 30 feet away, traveling faster than the Shinkansen "bullet train." As it passed, there was a 4-inch gap underneath it, and its passing made no more noise than a surfboard.

Those two experiences are from the world of magnetic flight—a world that all of us could be entering by the beginning of the 1990s. Magnetic flight will create a worldwide market as important as that of the steam locomotive in the 19th century. Magnetic flight is a revolutionary technology that could bring about reliable, safe, very high-speed intercity transportation. It would use no petroleum, and its impact on the environment would be far less than that of cars and planes.

JAPANESE BULLET TRAINS

Conventional wheel-on-rail trains best serve densely populated urban corridors, so of all advanced nations, Japan, with its linear north–south axis of nearly continuous cities, is best suited to railroads. In Japan most distances are too short for efficient air service.

During the reconstruction of Japan after World War II, high priority was given to the upgrading of the railroads. Railroads in the United States had long since begun to decline following their peak in the late 1920s, but by 1964, Japan began scheduled service on an entirely new kind of railway train, the Shinkansen electric bullet train. Today's bullet trains operate on special heavy rails welded into continuous sections more than a mile long. The rails are separated by a broad track gauge of 2.3 meters (90½ inches). (U.S. railways have a track gauge of only 56½ inches.) The first Shinkansen line ran for just 300 miles, from Tokyo to Shin Osaka. Not until 1975 was it extended to the southern end of the main island of Honshu, and beyond through the 10-mile Shin Kammon Undersea Tunnel to Hakata, on the island of Kyushu. Hikari superexpress trains now make the 1,070-km run (the distance from Portland, Maine, to Richmond, Virginia) in a scheduled time of 6 hours 40 minutes, running at a maximum speed of 210 km/hr (130 mph). Between 6:00 A.M. and 7:00 A.M. alone, seven Shinkansen trains depart Tokyo station for the south, and the last Hikari to make the full journey does not leave until 5:00 P.M. Every day about 120 Shinkansen runs are made in each direction, many of them over distances less than the full 1,070 km.

The bullet trains operate entirely under computer control, with the "engineers" normally simply monitoring displays in the train cabs. Even the track switches are set by remote control from a central control room in Tokyo. Since 1964, more than 1.6 billion passengers have ridden the Shinkansens without a single fatality. But the fares are not low; a ticket on the Shinkansen costs about as much as an airline ticket for the same route. Disregarding the huge costs (about $5.7 billion) for the original construction of the Shinkansen lines, the bullet trains now earn about $700 million per year, which is 75 percent more than their operating expenses. That helps offset the losses that JNR sustains on its conventional railroads. Overall, JNR, like nearly all railway systems, operates in the red and is sustained by government subsidies.

129

The number of riders on the Shinkansen peaked in 1975, shortly after the worldwide "oil shock" in 1973–1974, and has declined slightly since. But the bullet train is still very popular, and JNR is extending the Shinkansen line in several directions to the north. In 1982, new lines were opened from Omiya, near Tokyo, to Niigata, on the west coast, and to Morioka, near the north end of Honshu. The Tsugaru tunnel, 53 km (33 miles) long, is under construction to extend the Morioka line deep under the strait that separates Honshu from the northern island of Hokkaido. When that line is completed, Tokyo will be linked by Shinkansen trains to the resort area of Sapporo. Twelve additional Shinkansen lines have been proposed, and several are being surveyed.

While the bullet trains have been very successful in Japan, they push a technology—wheel on rail—about as far as it can go. The pounding of steel on steel as each Shinkansen goes by is so noisy that JNR has had to install sound barriers alongside much of the Tokyo–Hakata line, wherever it runs through densely populated areas. Considerable research effort is devoted to the problems that the Skinkansens will encounter as they speed northward into the snow and cold of Hokkaido; JNR may have to install snow melters along the right-of-way to allow the trains to operate at full speed.

A close look at the Shinkansen schedule reveals a problem that is already severe at 130 mph: shifting of the rails under the impact of the trains' passage. No Shinkansen departs before 6:00 A.M., and every one must complete its journey before midnight. Every scheduled Shinkansen carries recording devices to record jolts and bumps, and a special seven-car inspection train equipped with sensors and computers is run every week on the Hikari schedule, and every three months on the schedule of the slower Kodama trains. Each day the track is inspected by a patrol on foot, and every night, after the last Shinkansen goes by, a brigade of men and equipment swarms over the tracks to realign them according to the data from the central computer in Tokyo, before the first train rolls the next morning.

Above the Shinkansens' normal maximum operating speed, both track damage and energy costs rise very rapidly. In 1979, a Shinkansen established a world speed record of 198 mph; but in the 25-mph interval from 125 to 150 mph, the energy costs for all such wheel-on-rail trains nearly double, and track-maintenance costs increase by an even higher ratio, even though the rails are laid on concrete slabs for strength and durability.

THE TGV

Gallic pride surely played a part in the design of the only train that regularly operates faster than the Shinkansens. It is the French TGV (*train à grande vitesse*), which runs at 165 mph over a special line completed in 1981 at a cost of $1.6 billion, to cover the 264 miles between Paris and Lyon. The TGV upset the earlier Japanese record by hitting a peak speed of 238 mph. Lyon is the gastronomic capital of France, and French gourmets relish the fact that the TGV can whisk them from the capital to such legendary restaurants as La Pyramide in 2 hours and 40 minutes. (The eventual goal is 2 hours.) But on a typical weekday morning, the first-class cars on the TGV are filled with businessmen, scanning computer printouts rather than menus.

POTENTIAL APPLICATIONS IN THE UNITED STATES

For the travel conditions of Europe and Japan, trains running at 130 to 165 mph attract large numbers of passengers. But in the United States, competition from automobiles and commercial aircraft is greater. Also, the speed limits set for new trains of the Shinkansen or TGV design as a result of environmental impact and the costs of energy and maintenance make them unattractive to potential passengers in the United States.

The Budd Company of Philadelphia, long a manufacturer of railway equipment, has completed a survey for a proposed railroad linking Los Angeles to the pleasure palaces of Las Vegas, Nevada. The distance to be covered is roughly the same as from Paris to Lyon. By car it takes six hours. Pacific Southwest Airways has a sign near the highway just at the halfway point, in the middle of a monotonous desert. On it weary drivers read that PSA could have flown them to Las Vegas in fifty-nine minutes. The studies carried out by the Budd Company showed that all the economic considerations for the planning of a rail line—costs for the right-of-way, for construction and for equipment—were unimportant compared with the single issue of "ridership." A 250-mph travel speed would be enough to make drivers forsake their cars. A speed of 150 mph would not, and a train operating at that velocity would attract too few riders even to pay its operating costs.

For shorter journeys, trains in the Shinkansen speed range may compete more successfully. The American High Speed Rail Association, a private group which has borrowed $750,000 from Amtrak and raised $5 million from a Japanese financier to do a feasibility study for a Los Angeles–San Diego rail line, estimates a $2-billion construction cost for that 130-mile stretch, and hopes that its projected one-hour travel time for nonstop express trains will be short enough to pull drivers out of their cars. Studies are also being carried out for the routes Houston–Dallas, Chicago–Milwaukee and Miami to Orlando, Disney World and Tampa. In the Northeast Corridor from Boston to Washington, conventional rail traffic is heavy, and a track-improvement program costing $2.2 billion is due for completion in 1985. When it is done, trains will be able to attain 120 mph over that route. Such a speed, already routine in the mid-1930s for the streamliners of the Union Pacific, Santa Fe and Southern Pacific railroads, is about the safe limit for trains built to the old-fashioned track gauge.

Since 1974, Ohio has looked toward an eventual Midwest Corridor linking the state to Pennsylvania on the east and to Indiana, Michigan and Illinois on the west. State Representative Arthur Wilkowsky of Ohio, often called the father of the Midwest Corridor project, wants to see the states carry out the $10-billion venture on their own: "We don't want Amtrak service. Its record is terrible. We intend to build our system on the ashes of Amtrak." Robert J. Casey, Executive Director of the Ohio Rail Transport Authority, joins Wilkowsky in the view that Ohio needs a new railway line in the Shinkansen speed range or higher. But he recognizes the magnitude of the task. Says Casey, "It took a presidential directive by John F. Kennedy to commit the nation to reaching the Moon within one decade. In Columbus, we say, 'By the end of the decade, we'll reach Cleveland.' "

MAGNETIC FLIGHT

Two of the most serious limits to the speed of railroad trains—track-maintenance costs and noise—both result from the extreme shock forces of steel wheels pounding steel rails. Typically, a load of more than 7 tons is carried by each wheel, and must be transferred through a few square millimeters of contact area to the rail below. Only the bending of the steel itself keeps the resulting pressures lower than the breaking point, and in time fatigue cracks develop in the rails and wheels, forcing their replacement on a regular schedule. As speeds increase, track alignment must be held within even tighter tolerances to keep heave and sway motions within tolerable acceleration limits for the passengers.

For more than a century, railway engineers have recognized that they could relieve those problems if the weight of the moving train could be transferred to the earth below not through a few concentrated points of direct contact but over a large area, through some distributed, resilient lifting force. In 1864, an experimental vehicle was built that glided on a layer of water confined in a trough, and propelled itself by water jets. In 1889, visitors to a World Trade Fair in Paris were able to ride a later model of that vehicle. But the water-sled "train" had its own speed limit. Water becomes a hard medium when contacted at high speed, because unlike air, it is not compressible.

Only much later was it realized that the softest, most resilient distributed suspension for high-speed trains would be magnetic. The basic scientific truths necessary for that engineering achievement had been found shortly after the Civil War. But almost fifty more years went by before engineers applied those truths even to an experimental vehicle. In 1868, James Clerk Maxwell devised four basic equations that describe all the interactions between electric charges, currents and the electrostatic and magnetic fields they produce. The Maxwell equations describe how motors, generators, the telephone, the telegraph, radio and radar work—and even the inner workings of a modern microwave oven. Applied to travel, they describe the forces that act on one electric current when in the presence of another. Those forces can be very strong. When two cables each carrying 25,000 amperes run parallel, one an inch above the other, they attract each other with a force greater than half a ton for every meter of their length. If one current is reversed in direction relative to the other, the force is just as strong, but it becomes repulsive rather than attractive.

All magnetic-flight systems use one or the other of those two geometries. The two geometries correspond to the interactions between lines of bar magnets, each current being represented by a line of magnets. In the first case the magnets are arranged, from top to bottom, as north/south, north/south. The adjacent poles, one a south and the other a north, are opposites and attract. The second case, with the lower current reversed, corresponds to north/south,

south/north. The adjacent poles are both south, and repel each other.*

Supporting a train by the magnetic force depends on how that force varies with distance. If the separation between the centers of the currents is doubled, the force drops to half its value; if the separation is halved, the force doubles. That behavior of the force between currents has led engineers in two directions, and the debate over which is better continues.

Magnetic flight can be achieved either by the use of attractive forces, based on currents flowing in the same direction, or by the use of oppositely directed currents, to achieve repulsion ("dynamic maglev"). The fundamental simplicity of both solutions is masked, because in both cases the currents in the supporting rails are not flowing in cables connected to generators. In attractive maglev, electromagnets in the train pull upward toward steel rails. Those electromagnets are equivalent to a line of bar magnets, and they induce a corresponding top line of bar magnets in the steel rails by the magnetization of the steel itself. A miniature demonstration of the same effect occurs when a magnet is held below a steel plate that is anchored to a table. If the magnet is released, it will fly upward to the steel and stick to it. That demonstration also shows that attractive maglev is inherently unstable: the closer the magnet comes to the steel rail, the stronger is the force pulling it still closer. To take that unstable force and out of it produce a stable mechanical situation, with the electromagnets suspended at a fixed distance below the rails, the engineers install sensors that constantly measure the gap, and feed back amplified information electronically to control the electromagnets. If the gap becomes too small, the sensors feed back a signal that reduces the current, weakening the pull. It would seem risky to use an inherently unstable force to provide the lift, but modern solid-state electronics is so reliable, and can so easily be duplicated in parallel for redundancy, that the attractive-maglev system has proved to be safe in practice.

In dynamic maglev, the train supports itself above a metal

* Shortly before World War I a French inventor, Émile Bachelet, built and demonstrated a magnetic-levitation transport system at an exposition in Paris. Then in the early 1930s, Edwin Northrup, at that time a professor of physics at Princeton University, built a magnetic-flight system of his own. In Germany, magnetic-flight research was further advanced in that decade by the work of Hermann Kemper, who is regarded there as the "father of the electromagnetic-levitation train." As electromagnetic levitation becomes a mature field of engineering, the phrase is usually shortened to "maglev."

"guideway" by inducing currents in it. That induction effect, predicted by Maxwell's equations, is fundamentally stable. The induced currents always flow oppositely to those of the actual cables mounted in the train. And if the train sinks lower toward the guideway, the repulsive force increases. But for dynamic maglev also there is a problem: the induced currents flow only as long as the cable currents above them are moving. When the train stops, the induced currents die away and the lifting force is lost. In that respect, dynamic maglev is like aircraft flight; like airplanes, dynamic-maglev trains must carry landing wheels, which they can retract only when they have achieved "flying speed." Engineers concluded that dynamic maglev could be practical only if the currents flowing in the train's cables were very strong, so that the flotation distance could be several inches, compared with a typical gap of less than half an inch in the case of attractive maglev. To obtain those strong currents without using too much power, one must go to superconductors, and that requires the continuous operation of refrigerators on board the train, able to cool special alloys (niobium-tin is a favorite) down to temperatures only a few degrees above absolute zero. No large application of cryogenics (low-temperature technology) has ever been made commercially, so potential buyers regard dynamic maglev as a longer-term, higher-risk alternative to attractive maglev. But with its large clearance of several inches from the guideway, the dynamic-maglev train can attain very high speeds without risk of direct contact. Both the attractive- and dynamic-maglev trains are accelerated and slowed down in the same way, by magnetic forces produced between steady transverse currents in the vehicles and controlled, time-varying currents in the track or guideway. That arrangement is called a "linear motor," and its basic principle is the force between parallel currents, the same principle used to provide lift.

The history of magnetic flight in the United States is discouraging. While Edwin Northrup had built successful test models of maglev systems in the 1930s, no American scientist was able to obtain support for maglev research until 1963. Then a program to explore high-speed ground-transport systems was begun with modest funding from the Department of Transportation (DOT). For a few years the DOT maintained a test facility in Pueblo, Colorado, where it investigated several concepts for high-speed surface transport. The Tracked Air Cushion Vehicle (TACV), which was a

20th-century version of the French water sled of 1864, using compressible air rather than incompressible water as the supporting fluid, was tested and rejected because it required excessively close tolerances on its guideway. Dynamic maglev was explored through studies by SRI International and by a consortium made up of the Raytheon Corporation and of the National Magnet Laboratory, located on the campus of M.I.T.

But early in the 1970s, the DOT cut off all U.S. research on high-speed ground transport and shut down the Pueblo test facility. The work at Cambridge, directed by Dr. H. H. Kolm, continued for a short time on funding from the National Science Foundation, but despite successful tests on scale models, that work too had to be shut down, when the Foundation's own budget was reduced. Ironically, American maglev research was terminated just when the United States first got the message that it needed a transport system independent of petroleum—at the time of the first major oil crisis, of 1973–1974.

GERMAN RESEARCH

West Germany continued a careful, well-planned research program aimed at world leadership. Nearly all that work was funded by the German Federal Government. Its motives were to generate a large new export industry and to stimulate existing German industries active in electronics and aerospace. Two of the technology leaders in Germany, Siemens Electric and Messerschmitt-Boelkow-Blohm (MBB), are particularly active in that program. (Many Americans have already ridden on MBB equipment without knowing it; MBB is a major partner in building the successful A-300 Airbus airliner.) Germany recognizes that no nation can expect a major industry to be born overnight, without careful nurturing. Says Gert von Lieres of MBB: "You can't start the development when you're ready to sell the new system—you have to start twenty years earlier."

Indeed, the German program picked up in the mid-1960s where Hermann Kemper had left off. By 1969, Krauss-Maffei, a locomotive-building company, had built and tested Transrapid-01, a scale model based on attractive maglev with a linear motor. For the first several years of the German program, a competition of ideas was encouraged. Krauss-Maffei, experienced in building railway equipment, competed with MBB, whose main expertise was in sophisticated control systems and lightweight aerospace vehicles. Many of the MBB engineers working on maglev were from its space-flight division. In 1974, MBB and Krauss-Maffei joined their efforts in a partnership known as Transrapid-EMS.

To hold down costs for a commercial system, they decided to keep the supporting rails very simple—no more than plain steel sheets. That raised a potential problem, because in solid sheets induced currents would flow when the electromagnets of the fast-moving vehicle approached. That would give an unwanted dynamic-maglev effect (repulsion) partially counteracting the lifting force produced by attractive maglev. Once the first few electromagnets in the continuous line had passed, the steel rails would be thoroughly magnetized, and no further induced currents were expected to flow.

To evaluate the problem, Transrapid-EMS built the KOMET test vehicle, weighing 10 tons and designed to operate on a guideway just 1,300 meters (less than 1 mile) long. To obtain the high speeds that KOMET was designed for, in an acceleration distance of just 300 meters, KOMET was pushed by a sled driven by jets of hot water propelled by steam. After acceleration, the jet sled dropped off, KOMET coasted through a 300-meter test section and the last 700 meters of guideway was used for braking.

KOMET was put through its paces in 1976, and the results were good. At 200 kph (120 mph), only the first two of five lifting magnets along each side of the vehicle showed significant loss of lift, and at 400 kph (250 mph) the effect was no worse. On a long train it would be negligible. To make sure that a passenger-carrying vehicle would be fail-safe, the engineers subjected KOMET to stringent tests, one of which involved turning off one of its magnets when it was at full speed. The tests also revealed a side effect of induced currents: when the suspension magnets were rigidly attached to the vehicle, the power required for lift was much higher than when the magnets were held by springs so that they could follow the bends and twists of the rails. With the softer suspension, KOMET's magnets required only 30 kilowatts, or less than 4 horsepower, per ton of lift, even at the vehicle's top speed.

Thyssen joined the KOMET consortium, and as a result, Transrapid-05 was built, a full-size maglev system that transported 40,000 visitors during an International Transport Fair held in Hamburg in 1979. Transrapid-05 was a two-car train, with a total weight of 36 tons and a capacity of 68 people. Additional useful data came out of it, because its elevated guideway, designed to be assembled and torn down quickly, was so flimsy that the vehicle's magnet controls had to be made more complex to cope with the variations.

By 1978, with the KOMET data already in hand, the West German Federal Minister for Research and Technology (Bundesminister für Forschung und Technologie) decided to evaluate the maglev train under practical operating conditions on a full-scale test track. The train would have to run on schedule in all weathers, over a European temperature range of −25 to +40 degrees Celsius (−13 to +104 degrees Fahrenheit) in winds as high as 60 mph. Such tests, to be carried out over a period of years, would verify the practicality of maglev, and confirm that a maglev system would be safe, benign to the environment and comfortable for passengers. They would give data on maintenance and operating costs, so that by 1986 a fixed-price bid could be made for construction and operation of a revenue-earning maglev train line, whose operating costs would be paid by passenger fares.

The test facility would be expensive—its projected cost was DM 420 million, or about $210 million, for the first-phase segment, 20.6 km long. But that was, for comparison, only 3 percent of the subsidy that the U.S. Government was paying out every year to sustain the money-losing Amtrak. The money spent in Germany would be an investment, not a dole. As the Bundesminister said:

> A modern society stands or falls by the state of its research and technological development. In the Federal Republic, a special ministry takes care to ensure that important developments for the whole society are not allowed to become a matter of chance.

Seven German companies, each one of which had built up expertise in maglev technology over the previous nine years, joined to form Magnetbahn Transrapid, for the construction of the new facility and its maglev train. They were AEG-Telefunken, Brown-Boveri (a German branch, based at Mannheim, of a long-established Swiss firm), Dyckerhoff & Widmann, Krauss-Maffei, Messerschmitt-Boelkow-Blohm, Siemens and Thyssen Henschel. The project leader, Erich Eitlhuber of MBB, brought to what became the Emsland venture a background of twelve years in space-flight engineering. Given the very strong environmental concerns that prevailed in West Germany, the site had to be selected largely according to population density and the proximity of nature preserves. A site was chosen in the Weser-Ems district, close to the Dutch border at the northwest corner of Germany. No towns, villages or nature preserves were nearby, and the straight portion could be built next to the half-finished bed of the never-completed Gleesen–Papenburg canal.

A high-speed loop with a radius of just over 1 mile ran through flat farmland at the north end of the straight portion, redirecting the train to the south. And a low-speed loop at the south end was to run through forests and hilly terrain, where a maximum grade of 3.5 percent would be encountered. It is one of the advantages of magnetic flight that a maglev train, driven by the force of magnetism and making no contact with the rails, can climb steep grades in all weathers. Conventional wheel/rail trains depend on friction for their traction, and their load-hauling performance is much reduced by rain or snow. All but the south loop was completed early in 1983, with the remainder to be added in 1984–1985. The total length would be 31.5 km (about 20 miles), and in the Emsland there was room for additional length if needed later.

The name chosen was simply the Transrapid Test Facility in the Emsland (Transrapid-Versuchsanlage im Emsland, or TVE). Günter Steinmetz, an MBB aerospace engineer with hovercraft design experience, had worked on maglev since 1975 and was chosen to be project leader for magnetic levitation and guidance.

In a maglev system, about 75 percent of the total capital investment is in the guideway. That tends to push the design toward sophisticated vehicles which can tolerate a relatively crude guideway structure. For the TVE, the guideway is made up of beams 25 to 37 meters (82 to 123 feet) long. They deflect a quarter of an inch (6.3 mm) under the train's weight, and Steinmetz had to design his levitation system to cope with total guideway errors of as much as half an inch (12.1 mm) from an ideal straight line. By contrast, the conventional rails of the French TGV system must be held to far tighter (and correspondingly more expensive) tolerances of 1.2 mm (about a twentieth of an inch) over each 20-meter segment—error limits 10 times as stringent as those of the maglev guideway.

In other forms of transportation, passengers routinely tolerate accelerations and decelerations of one third of a gravity—enough, if sustained, to reach a 230-mph speed in half a minute. But no one asked for so much, so the train operates at a peak acceleration of less than a tenth of a gravity. The guideway costs are dominated by the structure necessary to take the much higher accelerations (0.37 gravity) of emergency braking, which is carried out by mechanical skids pressed hydraulically against the rails from both sides.

Transrapid-06, the maglev train designed to operate on the TVE, consists of two cars, carrying 196 passengers in five-across seating with a wide aisle. Each car has four double doors, so passengers can board or exit quickly at the station stops. Built largely of Space Age aluminum honeycomb sandwich panels, the cars weigh 61 tons each. Each car rests on pneumatic springs carried by four "bogeys," corresponding to the structures that hold the wheels on an ordinary railway car. The bogeys curve down and inward from each side to enclose the wide ribbon of guideway at its edges, preventing any derailment. Each bogey is suspended by ordinary steel springs connected to eight levitation magnets and seven guidance magnets which run just underneath the rails.

Steinmetz had to design his magnets for a mean time between failures of better than 9,000 hours, to achieve an overall reliability of .998 in an 18-hour day. The design was sufficiently redundant and fail-safe to enable him to meet those specifications even if 15 out of his 120 magnets failed. On the rare occasions when a magnet of the TR-06 does fail, the steel springs it is mounted on simply pull it down away from the rail, to a rest position on its bogey. Every one of the 120 magnets is controlled and powered independently, as part of a system that is decentralized to insure against the dominolike failure of a chain of components set off by a single breakdown. A high-frequency magnetic sensor continually measures the gap separation from magnet to rail, and controls the current to hold that gap constant. Ultimately, all 120 magnets draw their power from onboard storage batteries, of which there are two complete sets for redundancy.

Perhaps the most challenging problem that Steinmetz and his colleagues faced was keeping those batteries charged. The engineers rejected brushes, like the sliding overhead contactors of the Shinkansen or the TGV, because brushes were unreliable at the 400-kph speed they were aiming for. Their solution was to slot the rails of the guideway in a transverse direction, with slots an inch deep and wide, forming a kind of steel washboard. Windings were placed in the poles of the levitation magnets hanging below the rails, so that when the train was at speed those windings would be subjected to alternating magnetic fields as they sped by the washboard slots.

The principle was the same as that of an ordinary electric generator, and the result was the same. Steinmetz's "linear generator" produced electric current that could be used to keep the batteries charged.

The TR-06 train was so energy-efficient that its whole 122 tons could ease through the air at a total power of only 370 kw (under 500 hp), even at its maximum speed of 400 kph. Less than 60 percent of that power was needed for levitation and guidance; the rest overcame aerodynamic drag and ran the climate-control system needed to keep the passengers comfortable. Above 200 kph, the linear generators provided enough excess power to recharge the batteries, to compensate for the drain to which they were subjected when running heaters or air conditioners at the station stops.

Like monorail systems to which we have grown accustomed at airports and theme parks, the Emsland guideway runs 16 to 20 feet above ground level. It is supported by bearings on graceful A-shaped concrete pylons, and can be adjusted laterally by screw jacks. During construction, a course-tolerance framework carried an automated track-laying train which installed the rails to the fairly relaxed tolerances required.

For Krauss-Maffei, charged with responsibility for installing the rails, a major challenge was the high-speed switches at the ends of the straight section. The switches had to be safe for straight-ahead travel at the TR-06's full 400-kph top speed, and had to ease the train safely over to the alternate route at 200 kph. In Krauss-Maffei's solution, the rails of the guideway are elastically bent from a straight line to a curve of radius 2,200 meters (about a mile and a half) by gear motors spaced along the switches' 132-meter length. The process of resetting a switch takes twenty seconds.

Because the TVE is a research installation, it is instrumented even more thoroughly than a revenue maglev system would have to be. Switches are monitored and controlled by fiber-optic data links. A slotted waveguide for microwave transmissions runs the entire length of the guideway, and antennas on the train dip into the waveguide to send and receive a data stream equivalent to 60,000 simultaneous telephone conversations. Passive markers, sensed by the TR-06 as it goes by, permit continuous precise monitoring on board of the train's velocity. (Remember, there are no wheels turning whose revolutions could be counted.) Like all modern high-speed trains, the TR-06 is controlled entirely by a central computer, but its controls have many levels of redundancy, and its mechanical brakes are applied automatically if the onboard sensors detect over-speeding.

The seven companies who built the TVE are expert in design and manufacturing, not in operating a railroad. Magnetbahn Trans-rapid turned over the operational responsibility for the TR-06 to another consortium, made up of Lufthansa, the German Federal Railways and a consulting engineering firm which analyzes the test data. The test running goes on eighteen hours a day, with a six-hour break for maintenance and inspection. Until the south loop is completed, each test cycle consists of an out-and-return covering a total distance of 28 km, with a peak speed of 280 km (175 mph) reached on the traversal of the north loop. When the line is completed, the TR-06 will leave the test center near the south end of the 7-km straight section every half hour and in the next twenty minutes will make two round trips totaling 78 km. In each standard twenty-minute test run, the train will reach 400 kph (250 mph) over a length of more than 1 kilometer on the second of three traversals of the straight section. Slowing to a stop after the second passage through the north loop, the train will pause at the test center for ten minutes out of every half hour, for inspection and passenger boarding. If the TR-06 runs with full passenger loads throughout the planned ten-year life of the TVE, with the 90-percent availability that is planned for it, by 1993 it will have carried more than 17 million people.

The TVE, realistic as its operations are, is a research facility rather than a here-to-there transportation line for fare-paying passengers. The next step, its proponents argue, is likely to be a point-to-point connection between two major cities, or between a city and its most distant airport. (In Japan, JNR is building a special Shinkansen line to cut down the two hours by road that passengers must now spend going from Tokyo to the city's Narita International Airport.) Ultimately, a 400-kph maglev system could link all the major cities of Europe.

THE ECONOMICS OF MAGLEV

Europe is a prime candidate for maglev transport because of its high population density, the relatively short distances between its cities and its vulnerability to a cutoff in oil from the Mideast and the Soviet Union. Dr. Andreas von Bülow, the West German Minister for Research and Technology, points out that almost 5 percent of the total area of his country is already occupied by highways, parking lots, railway lines and airports. Surveys have shown that more than half the German population lives close enough to some highway to be disturbed by its noise. And pollution from the oxides and hydrides of carbon, nitrogen and sulfur ruins all farmland in strips 150 feet wide bordering every major highway. By contrast, a maglev train, like any electric train, produces no pollution except the well-monitored, controlled emissions of a distant power plant. And a maglev train going by at 155 mph makes hardly more noise than a car passing at 60 mph. A TR-06–type maglev train at 155 mph makes only a tenth as much noise as a conventional wheel-on-rail train like the TGV at the same speed. The total sound energy made by the "whoosh" of a maglev going by at its maximum 400 kph is about the same as the sound from a TGV at half that speed, and is far less irritating to those who live near the guideway than the TGV's pounding of steel wheels on steel rails.

Lufthansa is interested in maglev development because the airline's flights are unprofitable over the short distances between European cities. It would gladly turn over that market to a fast surface transport system, and in that market a maglev system should compete successfully. Even an advanced conventional train like the TGV can beat the airlines in actual door-to-door travel time for intercity distances up to about 130 miles, and on the few routes where such trains run, the airlines have lost up to two thirds of their passengers to the railroads. The maglev will beat the airliner in door-to-door travel time for distances up to 650 km, and will be only a few minutes slower out to 900 km. Starting in Frankfurt, a maglev network of that range would cover all of Germany and France and reach as far down into Italy as Rome. And the ticket price is projected to be far below an air fare or the total operating costs of an automobile.

Maglev trains, running on electricity that can be produced from any one of a number of fuels, would give Europe a degree of energy independence that it could never have for its petroleum-driven cars and airliners. A single takeoff by a commercial short-haul aircraft uses more fuel than an entire German provincial town runs through in the course of a day. The TR-06 at 400 kph uses only about as much energy per passenger-mile as a car driven at a third of that speed, and it uses only half as much energy as a short-haul airliner.

The Bundesminister skirts a delicate area when he compares the relative costs of maglev travel with those of conventional slow-speed railroads. Like the railroads of most countries, the West German Bundesbahn is nationalized, and its many employees are entrenched in what Dr. von Bülow carefully refers to as "mature social and personnel structures." Two thirds of the Bundesbahn's total expenses goes to personnel salaries, and of that figure, 60 percent is for employees who work in administrative offices, in stations or on the trains. The remaining 40 percent of salaries goes into maintenance. The Bundesminister looks forward to the opportunity of making a fresh start with a highly automated maglev system that will have few moving parts, entail no direct contact between train and track and require very little repair.

But for exactly that reason, German Federal Railways has been less than wholehearted in its support for a maglev. As with the introduction of any new technology, those officials and employees who might be displaced by the change view it more as a threat than as an opportunity. Yet their fears may not be wholly justified. If the many surveys made for prospective European maglev systems are correct, the 400-kph magnetic-flight vehicles will prove so popular with travelers that the Bundesbahn will expand rather than contract, and with a far larger volume of business it should be able to keep all its personnel.

The Bundesminister looks toward an eventual network of guideways that could serve a population of 60 million people living in eight countries. Extending as far north as Glasgow and Edinburgh in Scotland, it would link the great industrial cities of Great Britain's Midlands with London, then cross to France through a general-purpose Channel Tunnel that is already under construction. Paris, Lyon, Marseille and Lille on the west would be joined to Milan, Turin, Florence, Bologna and Rome in the south. Munich, Hannover and Hamburg would form its eastern boundary, and it would connect all the great cities of Germany, Holland, Belgium, Luxembourg and Switzerland in between. Just 5,300 km of guideway, hardly as long as a single U.S. transcontinental railway, would be sufficient. Its price would be roughly $50 billion, and that investment could be recovered from a few years' savings in the annual national subsidies for conventional railway operation.

U.S. ROUTES

By 1982, with the Emsland Test Facility in an advanced state of construction and much of the TR-06 train assembled, three companies of the consortium took the necessary next step: marketing the TR-06 technology abroad. MBB and Krauss-Maffei, both located near Munich, joined with Thyssen Henschel in the state of Kassel to form Transrapid International (TRI). TRI is a consulting engineering and marketing organization. It normally works in partnership with a local consulting firm in a target market, the local firm providing traffic data and TRI the maglev expertise. TRI has made presentations to the transportation departments of many U.S. states, having identified a number of corridors in the United States and Canada that seem ripe for maglev service. Some corridors chosen by TRI include Cincinnati–Columbus–Cleveland; Cleveland–Detroit–Chicago; San Diego–Los Angeles–San Francisco; Jacksonville–Tampa–Miami and a Texas loop connecting Dallas, Houston, San Antonio and Austin; and in Canada, Montreal–Ottawa. The most thoroughly studied routes, however, are from Los Angeles to Las Vegas and from Chicago to Milwaukee.

Las Vegas draws most of its income from tourists, and the city's authorities believe they can draw a much larger number of Los Angeles tourists if they can offer them a fast, cheap, comfortable way to make the trip. By 1982, the City of Las Vegas, with support from its Convention and Visitors Authority, its county and the State of Arizona, had collected study funds to be matched by a grant from the U.S. Government. The U.S. Department of Transportation, rousing itself slightly from a decade-long period of ignoring new technologies, provided the matching funds (a modest $150,000), and the study took place in 1982. Las Vegas chose the Budd Company to carry it out, in cooperation with the Bechtel Engineering Company, and Budd subcontracted part of the work to TRI. The Budd engineers considered high-speed conventional rail systems (the Washington–New York Metroliner, a Canadian diesel-powered light-rail train, the Japanese Shinkansen and the French TGV) and the "superspeed" maglev systems under development in Germany and Japan. They deemed JNR's dynamic-maglev system too advanced for construction in the 1980s, but concluded that the TVE/TR-06 attractive-maglev system was practical and worth examination.

In sharp contrast to the high-density corridors of Europe, the potential routes for the Los Angeles–Las Vegas line all run through desert areas with very few inhabitants. Much of that land is federally owned and would be available for a right-of-way. The 1982 study compared the advantages of an existing railway line (the

Union Pacific), an optimized shorter railway line with maximum grades of 3.5 percent, a highway route (Interstate 15) and the routes of several power-transmission lines.

A TR-06 maglev train could climb grades of 6 percent, so the engineers were able to lay out a maglev guideway route, just 230 miles long, that would run parallel to Interstate 15 for much of its length, then climb to the Cajon Pass and descend into the Los Angeles basin at its southeast corner, near Pomona. The best possible routing for a wheel-on-rail train would be 30 miles longer. Because the City of Los Angeles had not decided where its terminal should be, the engineers chose hypothetically the Ontario district, at the northeast corner of the basin. The route to Ontario led through Palmdale in the Mojave Desert. Using the exact parameters of the TVE/TR-06 maglev system, with a 400-kph (250-mph) speed, the study found the trip time to be one hour fifteen minutes from Las Vegas to Ontario, and about two hours (with long stretches of slower running) for a maglev connection all the way to Los Angeles International Airport. A 150-mph rail system would take two hours fifteen minutes for the Las Vegas–Ontario section alone.

The study analysts projected a lower operating cost for the maglev system, but thought its capital costs would be slightly higher. (That conclusion was reached because the maglev train would be nearly 100 mph faster than a TGV-type rail train. For equal speeds, the MBB engineers had calculated that a maglev guideway would cost only two thirds as much as a TGV railroad.) All the systems compared by the study would cost about the same overall (about $2 billion) and could be built in the same length of time. Whatever the technology chosen, the Budd engineers assumed that the trains would be built in the United States, under license.

One advantage of all electric train systems—their freedom from the need for oil as a fuel—would be shared by maglev and TGV systems alike. And for both systems, environmental-impact studies would take a full three years, so that neither alternative could be in operation before about 1990. But when the Budd analysts carried out their ridership surveys, to find out from current visitors to Las Vegas (now 4 million per year from the Los Angeles area) whether either system could pay its way, they found a dramatic difference: the 250-mph maglev train would earn more than enough to pay for its operations; the 150-mph TGV could not.

When the city checked the effect on revenues, excluding the maglev ticket receipts and any income from gaming, it found that the maglev would draw about $174 million per year in spending by new visitors—about twice as much as a TGV.

In the days of 5-percent interest rates, even the capital costs of a maglev system could have been paid off over a period of years by the sale of tickets. But in today's world there would have to be, as William Dickhart III, Budd's Marketing Manager, says, "some public involvement," probably in the form of tax deductions for investment in the maglev guideway. That is a far brighter prospect than now exists for conventional Amtrak-type rail systems. There is not a single passenger railroad in the United States today that even pays for its operations, much less its capitalization. There continue to be proposals for new wheel-on-rail systems (such as a Shinkansen line from San Diego to Los Angeles, proposed by Japanese financiers, and a TGV from Philadelphia to Atlantic City, proposed by a French group), but the only other potential maglev route in America that has been subjected to engineering and ridership analysis would run from Chicago to Milwaukee, a distance of 85 miles. Making stops at the many cities along that urban corridor, it would take the traveler from downtown Chicago to downtown Milwaukee in just thirty minutes.

DYNAMIC MAGLEV

Although the very idea of contactless magnetic levitation strikes must people as futuristic, the TR-06 technology is in fact relatively conservative. In the TR-06, the conductors that carry currents for levitation and propulsion are made of ordinary copper, and they run at the same temperatures as the windings of standard electric motors. The other form of magnetic levitation, the "dynamic" form, using superconductors running at cryogenic temperatures near absolute zero, is much more exotic, and further from realization than the TR-06 technology. Yet there are scientists and engineers who feel that it will prove to be the ultimate solution to the magnetic-flight challenge.

In magnetic-flight systems of the TR-06 variety, which derive their lift from the attraction of parallel currents running in ordinary copper conductors, the gap separation between the vehicle current and the guideway current must be kept fairly small to hold the levitation power within reasonable limits. (Recall that for given currents the levitation force is inversely proportional to the gap separation.) Critics of the attractive-maglev concept point to that small gap separation as a potential disadvantage: they argue that the levitation magnets will occasionally contact the guideway, even if briefly, and will produce noise and wear as a result.

Early in the 1970s, the Siemens Corporation convinced the West German Government that it should support the development of another technology, dynamic maglev, in order to provide an alternative should the problems of an attractive-maglev system prove insurmountable. In dynamic maglev, the levitating force depends on high currents flowing in superconductors kept at a temperature near absolute zero. No power is dissipated in the superconductors. The magnetic fields created by those fast-moving currents induce currents in a guideway, and the interaction of the superconductor currents and the induced currents gives a strong lifting force. It is so strong, in fact, that the dynamic-maglev vehicle "floats" about 10 cm (about 4 inches) above the guideway. From 1972 to 1979, dynamic-maglev research was pursued vigorously in Germany, funded by DM 100 million from the West German Government. Three companies took part in it: Siemens Electric, AEG Telefunken and Brown-Boveri. The consortium built a massive guideway in the form of a circular ring 280 m in diameter, banked at a 45-degree angle. On it a 20-ton vehicle reached 200 km/hr, the maximum possible speed that the guideway could stand without tearing apart under centrifugal forces.

Although the dynamic-maglev project met its design objectives, it lost favor in Germany when the "oil-shock" crisis of 1973–1974 forced the Ministry of Science and Technology to put a higher priority on conserving energy. Because a dynamic-maglev train needs a continuous flow of induced currents rather than the passive magnetization of steel to get its lift, it dissipates about 5 times as much power in its guideway as a TR-06 train does. At a speed of 500 kph, that power lost to the guideway is only 20 percent of the train's total power, the remaining 80 percent all going to overcome aerodynamic drag. The power losses to aerodynamic drag are equal for dynamic- and attractive-maglev trains. But for both types, aerodynamic drag at 400 kph is only half as large as at 500 kph, so at the lower speed a dynamic-maglev train uses significantly more total power than its attractive-maglev equivalent.

Late in 1977, the German Ministry decided that the possible

disadvantages of TR-06 technology were less worrisome than the energy costs of dynamic maglev and its reliance on superconductivity. The Ministry settled on 400 kph as a nominal cruising speed, and terminated support for the dynamic-maglev system. In 1980, the dynamic-maglev guideway at Erlangen was torn down, and the three companies that had built it joined the Transrapid consortium. (All such test facilities have short lives; the KOMET track and the JAL guideway in Tokyo shared the fate of the Erlangen ring after they were milked for test data.)

In Japan, the history of magnetic flight has been quite different. There dynamic maglev is being pursued as the preferred technology, largely because of the leadership of Yoshihiro Kyotani. A mechanical engineer, Kyotani has worked for the Japanese National Railway since 1948. The Shinkansen's enviable safety record attests in part to his design skills. When Kyotani was in charge of maintenance for all of JNR, he had to oversee the expensive repair program that keeps the Shinkansen lines operating. Seeing the track gangs realigning the rails every night, he felt there must be a better way for trains to run than on steel wheels. He had followed early theoretical work on dynamic maglev carried out by two American physicists, G. Powell and J. R. Danby, at the Brookhaven National Laboratory, and became convinced that maglev could be the "better way." In 1968, when he was appointed Director of Technical Development for JNR, he was in a position to turn his ideas into reality.

At first Kyotani met with disbelief, but he has been well supported by influential government leaders. His first $30 million in research financing was authorized by the Deputy Minister of Finance. Kyotani's development program cost Japan $60 million through 1982—considerably less than had been spent by then on maglev development in Germany. His total authorization is $100 million, and so far he has been able to obtain additional funding as he has met each design objective.

Kyotani's research facility near Miyazaki, with some eighty employees, has turned out a steady string of accomplishments. In 1978, a year after it was opened, 3 kilometers of its maglev guideway was completed, and a test vehicle reached 301 kph on it. In 1979, the ML-500 test vehicle, weighing 10 tons, set a maglev speed record of 517 kph—a little over 300 mph. In the same year, the ML-500 was run through a simulated tunnel to check for aerodynamic instabilities; none were found. A new body was attached to the ML-500 to house a helium refrigerator, and the vehicle was tested with a sealed, closed-cycle helium system—a necessity for any future practical dynamic-maglev train. The ML-500 had been designed to hug a guideway shaped like an inverted T, rather like a monorail, but in 1980, Kyotani converted the Miyazaki test track to a U shape, to make room for a train of conventional shape, without an awkward divider down its middle. By 1981, the lead car of the new train, the MLU-001, had lifted off the guideway and flown at 250 kph. In January of 1982, Kyotani was testing a two-car version of the MLU-001. The train was equipped with seats, but Kyotani was delaying as long as possible applying for the necessary regulatory approval to "fly" it with passengers, recognizing that he would be inundated with visitors if he did.

The complex components of a TR-06 attractive-maglev train are its controls and its mechanical suspension. The MLU-001's controls are simpler, but the Japanese machine does have mechanical complexity, in the form of retractable wheels both underneath and at the sides, and a helium refrigerator for every car. Yet Kyotani argues that a practical passenger-carrying maglev system will not be expensive. Neither the JAL attractive-maglev guideway nor the JNR dynamic guideway required any realignment over several years of testing, though both rested on Japan's unstable, shifting soil made up of volcanic ash.

In Japan, the capital costs for a new railway line are dominated by the costs of land, and Kyotani estimates that a direct Tokyo–Osaka maglev, running as straight as possible on a line requiring many tunnels and bridges, will be only 20 percent more expensive than an equivalent Shinkansen line. The aluminum coils that Kyotani requires for his guideway, to provide lift, lateral guidance and propulsion, are simple conductors cast into the concrete of the guideway itself. The cars, he estimates, will weigh only half as much as the cars of an equivalent Shinkansen train, because they will carry distributed rather than concentrated loads, and will not be subjected to pounding. The greatest uncertainties are in the helium refrigerator and the superconducting coils. But the cost of those components is only 10 percent of the total for the system. And there, Kyotani argues, prices can only go down, because the technology of commercial cryogenics is still in its infancy and much is still to be learned. He is not worried that JNR might run out of liquid helium, because Japan's liquid-air industry has a large surplus of capacity, and helium drawn from the atmosphere is one of its by-products. There will be plenty of it, he has calculated, even if JNR eventually builds an extensive network of dynamic-maglev guideways and a fleet of trains to fly them. Seven years is the earliest Yoshihiro Kyotani feels he could have a passenger-carrying dynamic-maglev system in routine service for JNR.*

For estimating the market potential in maglev technology, several hundred million dollars' worth of development data have been provided by now by the work of the Germans and the Japanese. Clearly, those nations have earned the right to exploit the first opportunities for profit from maglev. The United States, by contrast, has missed its chance. A vigorous program equal to that of the Germans and Japanese combined would have cost us no more than 4 cents out of every hundred tax dollars. For that saving, we gave up a market potential of several hundred billion dollars worldwide. Maglev train systems of the TR-06 level are well worth our investment—but as buyers, not sellers. We have nothing to sell.

As a buyer, the United States, like any other advanced nation, would be purchasing the licensing rights to construct a maglev system. The site preparation and guideway construction would be done by Americans working for U.S. firms, just as we build our highways. In many respects, building a maglev network for the United States would be comparable to the construction of our interstate highway system. That too was of large scale and expensive ($160 billion in 1950s dollars), and it benefited the nation as a whole by making automobile and truck transport much faster and safer. If we laced the major urban corridors of the United States with maglev lines, the total length of guideway required would be roughly 4,000 miles. At the prices estimated for the TVE/TR-06 maglev system, the total investment needed would be about $26 billion—far less than the cost of interstate highways, and comparable to what we now spend in four years to subsidize conventional railroads. The nation would benefit through a substantially reduced dependence on foreign oil, through higher speeds and lower costs for passenger traffic over the most heavily traveled corridors, and through a reduction of pollution in precisely those urban areas

* Not, perhaps, unreasonably optimistic when one remembers that the entire Apollo Project was done in eight.

where it is now most serious. If successful, a maglev system could be made to cover the country through transcontinental links, an extension of the San Diego–San Francisco line to Seattle and a radiating network of feeder lines from each major city. We would end up traveling faster, cheaper, safer and with less environmental damage than we do now.

There is one way that we could do more in maglev than buy imported technology. That opportunity for a more active and profitable role lies in leapfrogging the technology level, not only of the TR-06 but of the MLU-001. "First-generation" maglev trains would cruise at about half the speed of subsonic airliners, and might dominate the market in high-traffic-density corridors for intercity distances up to more than 600 km. There would be high payoff in a maglev system of twice that speed—trains traveling as fast as or faster than subsonic jetliners. It would be impractical to attempt such a project for a train moving in the atmosphere, because noise and energy costs for 600 mph at sea level would be unacceptable. But in vacuum, the natural environment for maglev, there would be no noise, no sonic speed limit and no energy penalty for higher speeds. Small-diameter tunnels or pipelines, located underground, need not be expensive. And vehicles traveling very fast need not be of large diameter to transport large numbers of people and goods. Slim vehicles with two-across seating, traveling in vacuum at 1,200 mph at one-minute intervals, could transport far more people between cities like New York and Chicago than a fleet of 100 jetliners costing several billion dollars. And they would do so at about 5 percent of the energy cost of airliners, providing travel times of half an hour from city center to city center. I call such vehicles "floaters."

The technology to build them is already here, and American aerospace firms could do the job, just as the aerospace companies of West Germany have led the TVE project. As for operations, the airline industry has the expertise, the high-technology orientation, the traffic data and the marketing skills to sell traffic by the floater technology. It has already been shown the way by Japan's private airline JAL and by Germany's Lufthansa, both of which understand how much they could benefit by opening a whole new intercity market based on magnetic flight. As of 1983, the U.S. airlines were in trouble, brought about largely by high costs for petroleum and the maintenance of expensive airliners. But their skills remain.

It would be ironic if, after all their problems with fuel costs and public objections to noise and pollution in the 1980s, the airlines of America were to solve all those problems at one stroke by going underground in the 1990s.

CHAPTER 6

THE 340-MPH LIMOUSINE

While magnetic-flight systems can link isolated major cities and provide high-speed transport through densely populated urban corridors, they cannot serve most smaller towns. Nor can the commercial airlines, without losing money. Yet many travelers journeying for business or private reasons begin or end their trips in such towns or in rural areas. The only means of transportation that can provide fast, direct, economical service for those travelers is the private airplane.

We will take a hard look at the market opportunities that will open for private aircraft, and will find it to be, potentially, in the tens of billions of dollars per year. If that opportunity is neglected for too long by the United States, it will certainly be lost to our industrial competitors, primarily Japan and Western Europe.

The United States has already lost a substantial share of the market for commercial and executive aircraft. The Airbus Industrie A-300, a 250-passenger twin-jet made by a consortium of West European firms, has outsold competing U.S. models particularly in Asia, now the fastest-growing market for commercial aircraft in the world.

Japan's Mitsubishi Heavy Industries, in World War II the manufacturer of the Zero fighter plane, has been selling its MU-2 twin-propjet executive aircraft in the United States for more than a decade. One of the few new business aircraft to enter the market in the "depression" year 1982 was Mitsubishi's Diamond I, a nine-passenger twin-jet with very high fuel efficiency. Just as the dollar volume of private-car production far exceeds that for buses, the long-term potential market for private aircraft in a mature aviation industry should far exceed the market for commercial and executive aircraft.

From the beginning that market will be strong in the United States, Australia and the affluent regions of the Middle East, Africa and Asia. As regulatory progress is made, it should strengthen in Western Europe, where private aviation is now severely discouraged. It is unlikely ever to be significant in Japan, because of that nation's high population density and short distances.

We are used to thinking of private aviation as available to only a small minority of the population. It is hard to believe that it could ever become a means of transportation as ready and easy to use as the private car. Yet the chance is there. First, we must understand that—contrary to most Americans' impression—the light aircraft is efficient as business equipment.

Flying to Mojave from the San Francisco Bay Area, one April day, reminded me of both the esthetic and the practical reasons that private aviation is potentially a major growth industry. My flight was to the Rutan Aircraft Factory—"RAF"—at Mojave Airport. No airline, not even a commuter, served Mojave. To get there by commercial service I would have had to make reservations in advance, fly from San Francisco to Bakersfield, then rent a car and drive for almost two hours. As the day's business proceeded, I would have had to leave, whether or not my work was done, to drive back to Bakersfield in time to catch another commercial flight. As it was, I had taken off from Hayward, a small Bay Area airport; was flying directly to Mojave and would park the plane only a few yards from the RAF hanger. I'd leave when our work was done, and it would take me only 1.7 hours to go or return.

Flight in a light aircraft at moderate altitude can offer a truly beautiful view of our world: peaceful, relaxing, silky-smooth. One is close enough to the ground to enjoy its details, yet not so close as to invade the privacy of those living below. Private flight gives freedom from the tensions and hassles of modern commercial air travel: none of the major-airport parking problems, no tyranny of airline schedules, no waiting in line for check-in, no security inspection or having to carry baggage through endless corridors. Also, in a light plane there is comfortable seating with plenty of legroom, instead of the sardine-tight packing that has become an airline standard.

Commercial air travel has changed over the past twenty years, and very much for the worse. Most of the changes have come about as responses to soaring fuel prices. In the older planes, like the 1960s-vintage propjet Convair-580, there was lots of legroom and the seats were thickly padded. Nearly every widebody designed for nine-across seating has now been reconfigured to squeeze ten across. Airlines have pushed the rows of seats closer together—to a 30-inch spacing on many aircraft, and to 29 inches on some charter flights. A decade ago, 33 or more was the norm. Under economic pressures, some airlines have quietly resorted to cancelling flights out of smaller cities five minutes before departure time when the passenger list indicates the flight won't make money. The carriers' responsibilities to passengers have now been left mainly to the discretion of individual airlines.

Many of the frustrations of today's commercial air travel had their counterparts in ground travel half a century ago. They sparked the revolution that pulled travelers out of trains and into private cars. In the case of very long flights between major cities, the large jet aircraft is at its best, with its economies of scale and speed. But a great deal of travel is of a different kind. It starts in a smaller city or suburban area, where most people live. If the flight is on business, its destination is likely to be a smaller town in one of the less populated states, where most of the newer and more profitable factories are being built. If the flight is for pleasure, its destination may be a coastal or mountain resort area. For those kinds of travel, commercial aviation cannot do a very good job. The routing can never be direct, but must go through one of the major interchange cities, like Chicago, Denver, Atlanta or Dallas. Chances are there will be long drives at one or both ends of the flight, because only the two hundred largest U.S. cities have any air service at all.

Several times there have been optimistic predictions of a rich new market in personal aircraft. Planes would become as inexpensive as automobiles, it was said, and every family would buy one. For many reasons, that never happened.

First, aircraft manufacturers never had the sales volume that would justify tooling investment for mass production. A car manufacturer might turn out a million cars per year, and on that volume could justify an investment of several hundred million dollars in mass-production tooling. For decades automobile engines have been made that way, almost without human labor, by specialized machines—a method that production experts call "hard automation." That has made car engines relatively cheap; but it is possible only because of mass production.

In contrast, a thousand planes per year of a given product line is large volume for an aircraft manufacturer. Their engines have to be made nearly by hand, so though aircraft engines are less complex than car engines, they cost ten times as much. That is the case with every component that an aircraft manufacturer must buy. Inspection costs raise prices even on those few aircraft components which can be mass-produced. Some light-aircraft alternators, for example, are identical to those made by Chrysler for cars. But because they must be inspected and certified for aviation use, those sold for aircraft cost three times as much. If automobiles had to be made and inspected in as labor-intensive a fashion as planes, they would probably cost about ten times as much as they do.

The second main barrier to growth in the light-aircraft market has been the antiquated nature of the air-navigation system. No one will buy an airplane for any reason except sport unless it can be used reliably in nearly all weathers. The family setting out for a vacation expects, just as does the business traveler, to be able to make a trip on schedule unless the weather is actually dangerous. Much of the time, a plane can operate safely when clouds form a solid overcast too low to fly under. In those conditions one has to fly by Instrument Flight Rules, "IFR"—first called "blind flying" in the 1920s, when its technique was developed. A part of it, determining the attitude of the airplane in roll, pitch and yaw, is relatively simple. It depends on gyroscopes carried on the plane, and those instruments have changed very little in the past fifty years.

But the second and more difficult part of the blind-flying problem is answering the question "Where am I?" Unfortunately, there has never been a single system that could tell with high precision where a plane was. A patchwork of different systems has been put together, each trying to answer under slightly different conditions that same basic question. Navigation is based on a network of radio transmitters called VORs, dotted irregularly about the United States at typical intervals of 50 to 100 miles. A receiver on board each aircraft detects signals from those transmitters and finds the direction of the aircraft from the VOR. The precision of that measurement is very poor: even when the transmitter and receiver are adjusted to meet federal legal limits of error, the position of the plane measured by the VOR can be wrong at a distance of 60 miles by as much as 5 miles.

The Federal Aviation Administration (FAA) measures aircraft position in order to satisfy its responsibility for control of traffic, but by a different system, a network of radar sets that depend on receiver-transmitters called "transponders" on each aircraft. That method is basically the same as the "Identification Friend or Foe" system developed in World War II. The precision with which it locates aircraft is not much better than in the days of the Battle of Britain, and barely an improvement over the VOR method. The traffic-control network saturates quickly, because radar cannot really tell where an individual plane is within a large block of air-space several miles long and wide and hundreds of feet thick. As a result, the skies are said to be "crowded" when there may be only one aircraft per thousand cubic miles of airspace over the United States.

In our present air-navigation system there has to be a different onboard instrument for each different navigational function: one to find the bearing from a VOR, another to find the distance from the VOR, another to measure altitude and still another to report that altitude to the nearest FAA air-traffic center. A separate instrument is needed to respond to the FAA radar, another to tell how well the plane is lined up for approach to a runway, another to tell the pilot whether he is above or below the approach path and others to tell him how close he is to the landing point. Communications radios are needed to talk with controllers over scratchy voice-radio party lines. Still another receiver is needed for beacon transmitters set up to supplement the VOR network. And finally, a completely inde-pendent, battery-operated unit is required, useless dead weight in normal flight, that is supposed to transmit a distress signal if the plane crashes. Altogether, a plane must carry eleven different on-board instruments to be fairly well equipped for flight in other than clear-weather conditions. Even with the weight saving brought by the solid-state-microchip revolution, that totals 50 to 100 pounds, not including weather radar.

The eleven instruments, ten of them electronic, cost fully half as much as a bare airplane. Their complexity has made learning to fly correspondingly expensive, because every new pilot who needs to use a plane in a practical way must learn to work with all eleven systems simultaneously, without making dangerous mistakes. As a measure of the costs introduced by complication, a pilot must have about 100 hours of dual instruction and about 200 hours of total flight time to begin to fly by instruments on his own. Yet for the most difficult actual piloting skill—landing a plane safely—just 5 hours in a "pinch-hitter course" is sufficient. No wonder that many qualified pilots who fly for the love of flight rather than by necessity have abandoned conventional aircraft entirely for inexpensive, low-power "ultralight" aircraft. Ultralights are the only aircraft that do not depend on navigational instruments and are not regulated by the Federal Government. In 1982, twice as many ultralights were sold as all conventional aircraft, even though their safety record was poor.

In the matter of air navigation and traffic control, the FAA and all who use the airspace are caught in much the same kind of trap that ensnared the U.S. postal system. In both cases, elaborate, labor-intensive technical systems were built up over many years, each being at the time of its inception the best solution possible. Each of those technical solutions has now been rendered largely obsolete by new technology (automated package sorting and electronic mail in the case of the Postal Service), but the agencies involved find it very difficult to change because they have large complements of civil servants to support the old systems, and because changing would require crossing interagency jurisdictional boundaries.

In the FAA's case, the new technology involves orbital satellites—but that is NASA's turf, not the FAA's. The effect was illustrated in 1982, when—fully twenty-five years into the satellite age—the FAA mandated still another two ground-based systems to be added to the already expensive cockpit. One, called MLS, for "Microwave Landing System," will improve, somewhat, precision on the approach to a landing. Another, TCAS, for "Threat Collision Avoidance System," warns aircraft that have TCAS equipment about the presence nearby of other TCAS-equipped airplanes. But as TCAS works by having each aircraft send out interrogation signals that trigger response signals from every aircraft nearby, it saturates just when it is needed most—when there are many aircraft in the same small volume of airspace. Altogether, when MLS and TCAS are added, pilots will be saddled with buying, learning and then using thirteen different instruments to try answering that one basic question: "Where's the plane?"

The existing air-navigation and traffic-control systems require elaborate networks of ground stations and a work force to match, and the taxes to support them are paid by everyone who flies, including airline passengers. Under peak operating conditions, the existing air-traffic system can handle fewer than 2,000 IFR aircraft over the entire United States. The cockpit crews of those airplanes total fewer than 3,000 people—yet the FAA itself has 45,000 employees. Over the years, the Federal Government has routinely shaved the corners of the law to pay the FAA's bills and divert aviation tax money from its intended purpose—improving flight safety.

For years there has been an "Airport Development Aid Program," with a corresponding trust fund established by Congress to pay for airport improvements out of aviation-fuel taxes. But successive administrations have left the trust fund unspent, to make the federal balance sheet look better. Every year, they have sought to divert the trust fund into paying FAA administrative costs.

Finally, in August of 1982, the government succeeded in tacking onto an unrelated Senate bill—by a parliamentary maneuver that permitted no debate—an authorization for a fifteen-year program to add more complexity to the existing air-navigation system and pay routine FAA office expenses out of the aviation trust fund. A month later, an 8-percent tax to support that program was added to the price of every airline ticket sold in the United States, and taxes were doubled on fuel sold for light aircraft.

In the first five years (1983–1987) of that program, more than $18 billion of trust-fund tax money will be spent. Of that $18 billion, less than 1 percent will go for weather services, although the great majority of aviation accidents are weather-related. Only 6 percent will go into research. About 20 percent will go to major air-carrier airports, and hardly 3 percent will be spent on the many thousands of reliever and light-aircraft airports on which private aviation depends. By far the lion's share of the money—almost $13 billion— will go into routine FAA office expenses and the expansion and maintenance of the existing navigation and traffic-control system.

Nonairline aviation has insufficient political power to curb such governmental wrongdoing. There are too few voters associated with it, since fewer than 1 percent of U.S. adults are pilots. But when and if personal aviation becomes available to the average citizen and families come to depend on it, the political system will respond.

The history of the automobile is a close parallel. When cars were expensive and rare, punitive laws restricted their use. But when Henry Ford brought out the Model T, a car so cheap that every ordinary family could afford one, the political climate changed almost overnight. Unreasonable laws were changed, a massive program of road building supported by fuel-tax money was begun and America took to the highways.

For a corresponding revolution in private flight, two barriers must fall: the high price of aircraft and the demanding complexity of instrument flying. For the first time we know how to level those barriers, and none too soon, because other events have made opening the skies to private traffic more necessary than ever before. One is the fuel economy of the light plane compared with the big jet. Small aircraft burn less fuel per passenger-mile than commercial jets (typically half as much) because they fly at slower speeds, where drag is less. Until 1973, that didn't matter. Before the oil-price rise of that year, the cost of fuel was so low that airlines hardly needed to consider it. Now, it accounts for more than 50 percent of the cost in many airline flights. Airline deregulation and the resulting cancellation of service to smaller cities came about largely because the airlines could no longer afford to buy fuel for those less profitable routes. If you visit the air terminal in a city like Fresno, California, you find only rows of abandoned ticket counters and the temporary signs of small commuter companies. And at those cities which do retain airline service, you have to put up with the delay and inconvenience of baggage inspection now that hijackings are a fact of life.

Already we see responses to these pressures. Large corporations can afford to buy expensive airplanes and hire pilots trained to fly IFR, so they're not forced to depend on commercial service. As a result, a booming market in large executive aircraft, mainly turbo-jets, held up far into the 1981–1982 economic depression. Those planes are flying corporate managers on company business into smaller airports no longer served by major airlines.

To reduce the high costs of light aircraft, we have a new weapon. The revolution in robotics is having its greatest impact not in mass production, where hard automation has long been the rule, but in short production runs. Using the old methods, it didn't pay to automate a production line that would turn out only one or two thousand identical items. But with robotics, a few general-purpose machines can manufacture a wide variety of products. To switch from one product to another, one needs to change only programs, not the tools themselves. That makes automation and its economies practical even for short runs, like those characteristic of aircraft. Visual inspection, once a tedious and expensive job, can now be done by a video camera linked to a pattern-recognizing minicomputer. The light-aircraft industry is not yet automated to a significant degree; but only through automation can it ever become a major growth industry.

Which nation will seize that opportunity? There is a disturbing parallel between light-aircraft production and the history of computer electronics. In the 1960s and 1970s, U.S. electronics manufacturers tried for short-term price advantages by relocating factories to areas with low labor costs. The Japanese automated instead—and ultimately swept the field.

GEOSTAR

The second barrier, the cost and complexity of instrument flight, will yield only to technologies that did not exist when the FAA began building our present air-traffic system. One is orbital satellites, and the other is high-speed computers. The system that combines those two technologies to give pilots on one simple, inexpensive cockpit display superior information on "Where am I?" is called Geostar. It is being developed privately, without a penny of government money, and aircraft constitute only a small fraction of its potential market. The Geostar Satellite System is based on work of my own, on which a U.S. Patent was issued in 1982. The Geostar Corporation was formed early in 1983, and a private stock offering of $500,000 for it was taken up fully in less than a month. A million-dollar stock offering was made a month later, and was also taken up very quickly. I have been granted a leave of absence from Princeton University to serve as President of the Geostar Corporation during its start-up.

Geostar will be a general-purpose, high-precision system for locating positions and exchanging high-value messages. In its minimal version, for coverage of North and South America, it will use four satellites, all in orbit above the Equator at fixed longitudes.* The

* As in the cases of earlier satellite systems, Geostar was proposed first as a service to the continental United States and surrounding coastal waters. Its extension to all of North and South America, and ultimately to the entire world, depends on international regulatory approvals.

ground station, in the eastern United States, will send an interrogation radio signal up to one of the satellites many times per second. The satellite will relay the signal in a broad beam covering the Americas. Geostar's only similarity to existing systems is that it will work through a transponder. When the transponder, which can be in a car, truck, boat or aircraft or hand-carried, receives the signal from the satellite, it will respond on another frequency with a binary sequence ("fingerprint") of pulses that uniquely identifies it. A telegraphic message previously keyed in on the transponder can be added to the same signal. The transponder will then ignore all further interrogations for a time depending on the needs of the user: a fraction of a second for aircraft in the landing phase of flight, up to many minutes for ships far from land. That dead time will prevent saturation of the single Geostar channel.

The transponder's pulse code will be received at the satellites and relayed by them to the ground station over tight beams, arriving as identical pulse sequences at different times. The computer at the ground station will identify the fingerprint, measure the times and from that timing information compute the longitude, latitude, altitude and time of response of the transponder. From its memory it will retrieve the corresponding data from the last previous response and so calculate the transponder's velocity in all three coordinates (ground track, ground speed and rate of climb). After computation, the ground station will send all the navigational information to the user as a radio pulse sequence, which will go by tight beam to one of the satellites and from there to the specific user. His or her transponder will identify the information by its own unique binary fingerprint, and will present the data—most simply, as coordinates, velocities and a telegraphic message—on a calculator-like liquid-crystal display. The delay time between the user's request and his receipt of the information will be approximately half a second, and Geostar will give him his position with precision of a few meters or better.

As the Geostar system is not drawing on tax money, it will be supported entirely by service fees, like a telephone system. When a transponder uses Geostar, the computer will therefore update the service charges associated with the transponder, for later billing. Because Geostar will be so efficient, those charges can be very small.

The best estimates made so far indicate that Geostar will require private investment of less than 1 percent of what we spend on the

FAA over a ten-year period. Yet Geostar will offer the pilot for the price of a small home computer far more complete and effective position information than can be obtained from the old ground-based systems even with a million-dollar cockpit. As the Geostar central computer, backed up by many levels of redundancy, will have all the information on every transponder-equipped aircraft, it can send that information to the FAA for the government's use in air-traffic control. In the event of an emergency, Geostar can guide the pilot to a safe landing on the nearest available level ground, even if that ground is shrouded in fog. It can also report instantly the location of any downed aircraft, greatly increasing the likelihood of saving the survivors. (In the FAA's ground-based system, typically twenty-four hours go by until a crashed aircraft is found, and many are never found at all.)

For routine flights, Geostar can provide to the pilot in the cockpit precise information not only on his own position but on the positions of all nearby Geostar-equipped aircraft, in every phase of flight. Unlike the federal TCAS system, Geostar does not saturate in high-density traffic areas. With Geostar's help the pilot can also make a precision approach in instrument conditions to every airport in the United States, no matter how small. Given Geostar's precise information on where to fly, there is no technical reason that its navigation data could not be routed directly to an autopilot, so that a plane could be flown safely from takeoff to landing by the central computer system, and kept clear of all other traffic en route.

It may be years before the jurisdictional and administrative underbrush is cleared away sufficiently to permit such flights, but when they do occur they will be a great deal safer than those now flown by human pilots. Then, for the first time, it will become possible to use a plane effectively and routinely without having to undergo the present demands and expenses of pilot training. With that final transition, the market for light aircraft can grow to equal the market for automobiles—several hundred billion dollars per year, worldwide. Fortunately, the capacity of even the first-generation Geostar system is so great that it can accommodate 100 times as many aircraft as there are in today's air fleet.

In March 1983, Geostar's application for radio frequencies and its application to launch four satellites were submitted to the Federal Communications Commission (FCC). By September of 1983, the corporation was testing an Earth-bound equivalent of its complete system. In the tests, transmitter-receivers emulating the Geostar satellites were located on high peaks in the Sierra Nevada mountains of California. A ground station equipped with a minicomputer communicated through those transmitter-receivers to transponders carried by hand and in cars, trucks and aircraft. Positions were calculated by the minicomputer and relayed to the transponders.

The Geostar Satellite System has so many potential user markets that its economic success does not depend on aviation. But when the Geostar satellites are in place, possibly as early as 1987, aircraft can begin to use Geostar for messages and position information supplementary to those of the FAA. With practical operating experience over a period of years we can gain assurance that Geostar is reliable and precise enough to guide aircraft. With that assurance, we will have achieved the second of the two necessary conditions for rapid growth of the light-aircraft industry.

SAFETY

Shifting the burden of flying from individual pilots to an electronic system backed up in every detail by levels of redundancy will improve flight safety, and with it the public acceptance of private flying. While nearly every child enjoys flying, about 15 percent of all adults now avoid it completely, refusing even to fly on a commercial airliner. Another 15 percent (approximately) are willing to take a commercial plane, but not a light aircraft.

In fact, very few aviation accidents are caused by mechanical failures. Aircraft must be inspected regularly to satisfy federal laws, and maintenance on them is "progressive," to prevent future breakdowns.* Planes are designed for long life, and many continue flying for twenty or even thirty years.

The great majority of light-aircraft accidents are related to weather, and most of those accidents are of the kind that flight guidance by Geostar should prevent. In conditions of marginal visibility, pilots without instrument training sometimes become disoriented and lose control of their planes or, if attempting to land, are

---

* Piston aircraft used only as personal or family vehicles must pass annual inspections. Planes carrying passengers for hire, and all turbine-powered aircraft, are inspected every 100 hours of operation.

unable to land safely because they have not practiced using radio approach aids. A small fraction of all accidents are caused by errors of judgment in committing the aircraft to conditions it is not equipped for: flying into thunderstorms (no aircraft, whatever its size, can survive the turbulence of a severe thunderstorm) or into rapid icing. As the Geostar central computers will maintain records updated minute by minute on the location of thunderstorms and the location of regions of severe icing, there is no reason for a Geostar-guided aircraft to fly into either of these dangerous conditions.

Despite the imperfections of today's air-traffic system, light aircraft have about the same accident fatality rate per hour of flight as commercial aircraft: over the ten-year period 1971–1981, commercial air carriers had 3.7 fatalities per 100,000 flight hours; noncommercial planes were only slightly higher at 4.0 fatalities per 100,000 hours. But as air carriers typically transport far more passengers per flight than do light aircraft, the same statistics indicate that the chance of accident per flight hour is much higher for noncommercial than for commercial flights. That is certainly to be expected in today's pre-Geostar flying. Airline flights are over standard routes, into the best-equipped airports in the country. Light-aircraft flights are made over routes that are rarely the same twice, and into thousands of small, poorly equipped airports.

AIRCRAFT DESIGN

Aerodynamics is a well-established field of engineering, so aircraft shapes of the future cannot be very different from those of today. The best prototypes that are already flying preview the kinds of aircraft available a few years from now. They are of two basic designs, each produced by extraordinary engineering talent.

The story of the first begins in the California high desert, at Mojave. Mojave Airport's broad, thick concrete runways date from World War II, when it was home to Air Force training squadrons. The taxiways are thinner than the runways, and through every crack the desert weeds have grown. At the north side of the field there is a long row of derelict 707s, and at the east side a row of ex–Air Force Hercules transports. All are for sale "as is," giving Mojave the flavor of a giant, neglected used-car lot. On the south ramp, the tiny hanger of Burt Rutan's RAF company is so inconspicuous that it could easily be overlooked. But the walls of its waiting room

are covered with framed certificates from the Fédération Internatio-
nale d'Aéronautique: world records for speed, distance and efficiency won by aircraft Burt Rutan has designed.

Rutan, born in 1944, won a national award for his B.S. thesis in
aero engineering at California Polytechnic State University in San
Luis Obispo. There he became fascinated with the idea of a plane
that would remain safe no matter how ineptly it was flown. He
zeroed in on the most common type of fatal accident, the stall-spin.
It happens when a pilot, usually without many hours of experience,
does not achieve sufficient speed on takeoff or maintain it on approach to a landing. At slow speed the wings give less lift. The pilot
feels his plane sinking under him and tries to compensate not as he
should, by lowering the nose and adding power to gain more speed,
but by pulling back on the controls to raise the nose of the plane.
Past a certain critical angle, the wing stalls: the airflow around it
separates into turbulent eddies and it stops lifting. The plane drops,
and usually begins to spin at the same time. Only if it is several
hundred feet up can it recover before it crashes. In conventional
aircraft, it is difficult to make a plane stallproof without unduly
limiting its range of control.* So Rutan looked into an unconven-
tional geometry, with the main wing aft and a smaller "canard"
wing forward. The canard had been known since the days of the
Wright brothers, and early in the 1960s it had been used successfully in a fighter plane, the Swedish Viggen.

A conventional plane balances on its wing, which supplies all its
lift. For stability the center of gravity must be in front of the wing,
so the tail surface provides a downward balancing force. When the
nose is raised so far that the wing stalls, the plane is left with no
support at all. In a canard, both the main wing and the canard wing
supply lift. By setting the canard wing at a slightly greater angle,
taking a bigger bite out of the airstream, Rutan designed a plane
whose canard wing would always stall before the main wing. When
the canard stalled, the nose would drop, preventing the main wing
from ever stalling.

After his design work on canards at Cal Poly, Rutan spent seven
years as a flight-test engineer at Edwards Air Force Base, in the

* Through research at NASA's Langley Research Center near Norfolk, Virginia, and at
aircraft companies, some progress has been made in recent years in making conventional
aircraft, if not absolutely stallproof, at least highly resistant to stalling.

high desert 15 miles east of Mojave. In his spare time he scaled down the Viggen to a size that he could build himself, mainly out of plywood. He called the result the Vari-Viggen, and when it had been flight-tested thoroughly, he sold its plans to aviation enthusiasts so they could build their own Vari-Viggens at home.

Rutan pioneered a new construction method for home-built aircraft with the VariEze, a two-seater rear-engined canard designed to be built in any garage. It could be made without expensive tools, out of rigid plastic foam cut with a hot wire, shaped with a kitchen knife and covered with a smooth skin of fiber-glass cloth soaked in epoxy resin. Several thousand sets of plans for the VariEze were sold, and hundreds of the planes are completed and flying. On just 100 hp the VariEze cruises at 180 mph and gets about 35 miles to a gallon of fuel.

Rutan also applied his newly learned expertise in fiber-glass construction to an unsolicited proposal he submitted to the NASA Ames Laboratory in Mountain View, California. There Robert T. Jones of NASA had designed a new type of supersonic transport, whose rigid "oblique wing" was pivoted at the fuselage. The straight wing would extend out symmetrically to left and right to give the plane excellent, docile handling qualities for takeoff and landing. At cruise altitude the wing could rotate by as much as 60 degrees, like a swimmer doing the crawl, with one arm forward and the other back. At that angle the wing would slice the air just as efficiently as the swept-back wing in all modern jets.

NASA wanted a flight-test version, but couldn't afford the multi-million-dollar proposals for it that aerospace companies had submitted. In 1975, Rutan offered to do a feasibility study for NASA for the outrageously small sum of $560. NASA accepted, RAF delivered a forty-page report and Rutan further offered to design a prototype for $11,000. In his words, NASA took that "out of the coffee money." Rutan's next bid was to build the plane, a man-carrying, subsonic twin-jet whose wing could rotate through the full 60-degree angle that Jones needed. RAF, aided by the Ames Development Company on Long Island, finished it in just over one year, and delivered the plane fueled and ready for flight early in 1979. The total bill to NASA for the entire program was only $231,000. The flight tests of the AD-1 were completed successfully by the NASA Dryden Research Center two years later.

In 1981, Rutan accepted a subcontract to design and test a

manned flight demonstrator, scaled to 62 percent of the dimensions of Fairchild-Republic's "Next Generation Trainer," a two-place, fuel-efficient turbofan trainer for the Air Force. He had no choice in the aerodynamic shape of the trainer, but the design of everything inside its skin of paint was left to him. When the trainer was finished, RAF carried out the entire flight-test program for it in six weeks, then completed a final report for Fairchild in just one week more. Some months later, in 1982, the Fairchild/Rutan trainer was chosen by the Air Force for a production run of up to 650 planes, to serve through the years 1987–1995.

Along with his contract work, Rutan continued to develop home-builder projects to pay the modest operating costs of RAF: the Long-Eze, a heavier, longer-range derivative of the VariEze, and then a canard motor glider. On the side, he worked on the Quickie, a single-seater micro-airplane designed to fly on just 18 hp. With increasing reputation from his remarkable track record of success, he was able to experiment with more new designs: the Predator, a plane designed for agricultural spraying, and the Grizzly, built purely to satisfy his own quest for knowledge. The Grizzly, with tandem wings joined at their tips, is equipped with huge flaps and can land at only 35 mph. Out of it came an idea Rutan patented, for a way that airliners could employ large flaps without having to carry drag-producing tracks for them. In the long line of Rutan record breakers, the ultimate is the Voyager, designed to be flown round-the-world nonstop.

Rutan has kept RAF a very small operation, in order to retain his freedom to explore the state of the art in aerodynamics. His only design intended for the commercial, factory-built aircraft market is the Defiant, a four-place canard with twin engines in line at the nose and tail. Beechcraft, in great secrecy, has built a prototype that is a twin with many of the features of the Defiant. Rutan and Brandt Goldsworthy are working together on the Beech project. The Avtek 400, a 6-passenger twin turboprop business aircraft, also uses the canard configuration. Its extraordinarily light empty weight of 2966 pounds is achieved by a composite airframe of Kevlar and Nomex honeycomb. Avtek projects a top speed of 425 mph, transcontinental range at 280 mph, a ceiling of 38,000 feet, and a climb rate of 5,000 feet per minute.

Now that fuel efficiency and easy maintenance have become the highest priorities in airliner design, a new generation of commercial jets has been brought out, among them the Boeing-757 and 767. All have just two engines, instead of three or four. In light aircraft the same evolution toward fewer engines is occurring, for the same reasons.

Formerly, in charter service twin-engined planes were used almost universally. Now designers are bringing out a new generation of big single-engined planes. One can always get lower drag in a single, with the engine profile in the silhouette of the fuselage, than in a conventional twin with its two engines and a fuselage side by side, all to be dragged through the air.

Paradoxically, the switch from twin to single-engined light planes will make flying safer as well as more economical. The rate of fatal accidents caused by engine failures is twice as high for twin-engined planes as for singles. Engine failures occur at least twice as often in twins; when they occur at takeoff or landing, a conventional piston twin is very difficult to control, and unless the pilot takes exactly the right action within seconds, the plane is likely to half-roll onto its back and dive into the ground; finally, even if everything is done right, a piston twin running on one engine can climb only very slowly if at all. By contrast, when the engine of a single quits, the pilot concentrates his attention on a safe landing, and usually walks away from it.

In the new generation of big singles, which are really executive-transport aircraft, the heaviest models are powered by turboprop engines. All those planes are pressurized, and all are based on existing twin-engine six-passenger designs. Comfort and legroom in them correspond to first class in airlines. Beech Aircraft has its "Lightning," based on a pressurized Baron 58P model and powered by a Garrett turbine engine of 865 hp. Cessna has announced its Model 184, of similar design.

Because a turbine engine costs at least ten times as much as a conventional piston engine, the new pressurized turboprops are beyond the economic reach of most individual buyers. But at least two light-aircraft companies are developing new pressurized, six-passenger singles that can take either piston or turboprop engines. Those, in combination with the new developments in automation and satellite navigation, can spark a revolution that will give all of us a new freedom in travel. Families as well as small companies will be able to buy and use fast, comfortable private aircraft. Each of the new planes approaches the ultimate in aerodynamic efficiency. Each is the brainchild of an outstanding aeronautical engineer. And although the two competing engineers worked from wholly different philosophies, their two planes are remarkably similar.

The two designers are Roy Lopresti of Mooney Aircraft and Jim Griswold of Piper. Both were boys in the 1930s, the grand era of flight; the DC-3 was making commercial aviation economically practical for the first time, and the China Clipper flying boats were crossing the ocean on regular schedules. Lopresti graduated from New York University at 20 with a B.S. in aeronautical engineering, and along the way captured the Chance-Vought Design Award. The Air Force ROTC program had helped pay his way through college, so when the Korean War broke out he was called to active duty, earned his Air Force wings and flew a tour of duty in Korea. The Air Force then sent him to its research center at Wright Field near Dayton, Ohio, to design fighter planes. After his service at Wright Field, Lopresti joined Grumman Aircraft on Long Island, and Grumman sent him to Cape Canaveral in 1963 as the race to the Moon began. He ended up a consulting pilot at the Cape, standing in for astronauts in many of the ground tests of Apollo.

Then in 1972, Lopresti, beginning to get bored with the Apollo program as it wound down, volunteered to move to Cleveland. Grumman had just bought a small aircraft company, American Aviation, which had a line of simple, inexpensive single-engined planes. But they had indifferent performance, and Lopresti was fascinated by the idea of speeding them up by applying his own greatest talent—sleek aerodynamics. In only six months he designed the Grumman American Tiger, and it was a success. The Tiger kept the simplicity of its predecessors, with fixed propeller and landing gear, but its aerodynamic drag was so low that it flew as fast as many complex, retractable-gear airplanes. Grumman sold many Tigers in the next few years.

Soon after the Tiger made headlines in aviation magazines, Lopresti got another job offer. For years Mooney Aviation had built planes that were a little faster than its competitors', but were too small and cramped inside to be comfortable for long trips. Mooney had gone out of business, but a fiercely loyal Mooney pilot, Bob Cummings, was in a position to do something about it. Cummings was Manager of the Manufacturing Division of Republic Steel, and his company wanted to diversify. There was a good deal of specialty high-strength steel in every Mooney, and Cummings sold Republic Steel on the idea of buying the defunct company, sprucing up the product line and going into business selling airplanes. Then Cummings invited five of the best people he knew to come to Kerrville, Texas, and restart Mooney. One of them was Lopresti. Roy was doubtful, but Cummings won him over with an offer he couldn't refuse: complete design freedom, as long as the result was a good product in the business sense. The new Mooney would have to have both performance and good looks. Lopresti went to Kerrville in 1973 as Chief Engineer of the refloated company, joining the four other men whom Cummings had picked. Ten years later, three of the five were still with Mooney.

The new managers felt that it would be too costly to put a new engine into the old Mooney airframe, so they asked Lopresti to try improving the plane by aerodynamics alone. Lopresti's aerodynamic cleanup of the old design raised the plane's top speed to better than 200 mph. To do that, Lopresti had to reduce the aerodynamic drag of the Mooney by fully 20 percent—a remarkable accomplishment in the case of a design already noted for sleekness. They called the new plane the 201, and when it was announced in 1976, it too, like the Grumman Tiger before it, was an instant success. The sales backlog when the old Mooney company had gone under had been fifteen planes. In one day, after the 201 was announced, the backlog went to two hundred. While the new Mooney still had a relatively small cabin, it was significantly faster than its competition, and that sold airplanes.

Roy Lopresti, at the age of 50, had never yet found the opportunity to design a plane from nose to tail and see it through to production. But after several years of success with the 201, his colleagues at Mooney and their backers at Republic Steel felt the company was strong enough to make that investment. The time was right, in the early 1980s, for a new concept in light aircraft: a plane designed to combine high speed with great fuel economy, room and comfort in its cabin, and extraordinary attention to safety. Lopresti planned to sell the new plane to air-taxi and charter operators, corporations and the many self-employed entrepreneurs who had started successful companies and found they could improve productivity by flying their own planes.

The new plane couldn't be radically different in appearance from the designs its buyers were familiar with, because that might hurt sales, and Mooney was betting its company on the new design. For all its success, Mooney was a small firm, with only about $30 million of annual business. It would have to spend $22 million to develop, test, fine-tune and certificate the new plane, and then get it into production. It would leave to innovators like Burt Rutan the development of canard designs, with the smaller wing forward and the main wing aft. Lopresti, once he was convinced that a conventional aircraft could equal the performance of a canard, would stick with the conventional arrangement. But from that point on, the new design would part company with its competition. For starters, Lopresti was looking for a plane whose lowest flying speed would be barely 20 percent of its top speed. With so great a ratio, the stall speed could be kept to a safe 60 mph while the top speed could be pushed to over 300 mph. That settled the name of the new plane: the smaller Mooney was the 201; the new one would be the M-301.

The 301 would have to fly at 25,000 feet to obtain high speed with great fuel economy. If that was to be its normal operating altitude, the plane would have to be pressurized and capable of getting up there in a hurry. Lopresti solved that problem in cooperation with the Avco-Lycoming Corporation of Williamsport, Pennsylvania, a leading manufacturer of aircraft piston engines. He chose one of Lycoming's best products, an air-cooled 350-hp engine.

Lopresti and the Lycoming engineers worked out a new package consisting of a big turbocharger, a device called an intercooler and power-takeoff shafts that would allow the six-cylinder engine to drive a number of auxiliary pumps and generators. The turbocharger was sized to compress the thin outside air at 25,000 feet to sea-level density, so the engine would develop full power. But compressing air like that would also heat it; the same principle is used in diesel engines to make the intake air so hot that oil will ignite in it. Lopresti wanted his engine to run comfortably cool, so Lycoming designed an unusually large intercooler, a radiator that would transfer the heat of the compressed intake air into the cold high-altitude airstream flowing by it. With that package, the big engine could develop its maximum rated power with no time limit set by heating, and could haul 2 tons of airplane up to 25,000 feet in only twenty minutes—about as fast as a commercial jet climbs.

The wing was next in Lopresti's thinking, and it was critical to his design concept. For minimum drag at high speed it would have to be thin and small in area. But he also wanted docile, slow-speed operation for takeoff and landing, a requirement that called for a wholly different wing of much larger area. He solved the problem the same way it is solved for commercial jets: by using "Fowler flaps"—large plates that would tuck into the wings when the plane was flying fast. On the pilot's command they would extend backward and down to increase the wing area by almost 30 percent for slow-speed flight. He wanted those Fowler flaps on 90 percent of the span of each wing, which left very little room for ailerons, the movable surfaces that allow planes to bank. Again he used a trick that was common on large jets but rare for smaller planes. When a 301 pilot turned his control yoke left or right to bank the plane, a "spoiler," or hinged plate, would come up into the airstream and kill part of one wing's lift. In combination with vestigial ailerons near the wing tips, that would provide the 301 pilot with twice the authority in banking that he would have on ordinary aircraft.

Once that was settled, Lopresti contracted with one of the world's great aerodynamicists, Professor Eppler of West Germany, to design a slim, efficient wing that would retain gentle handling characteristics even at low speed. To achieve in production the exact contours that Eppler called for, every 301 wing would be manufactured from the outside in. The aluminum skin of the wing would be sucked outward to conform to a female mold, and while it was held there by vacuum the internal ribs would be riveted to suit the exact profile needed.

Lopresti's quest for beauty in the airplane had now succeeded as far as the wing was concerned. Eppler's long, slim tapered wing was a perfect example of beauty in functional design. But in the fuselage of the 301, Lopresti had greater freedom to choose among shapes, all of equal aerodynamic efficiency.

He planned a roomy, comfortable cabin wide enough to accommodate an aisle. Entry to it would be through a large door just behind the left wing. Once that was settled and the cockpit design was firm, with wraparound windows like those of a Learjet, Lopresti sketched the profile and silhouette of the new plane. Aerodynamics dictated most of his decisions, and esthetics gave him the rest, as he arrived at a fuselage made up entirely of long, smooth, flowing curves, ending in a tail swept back at 45 degrees. Many of the radio antennas needed on the plane would be concealed inside fiber-glass sections of the fin and tail cone, to minimize drag. That curving shape would be expensive to build, but it was essential if the 301 was to become the classic of design that Lopresti and his colleagues hoped for.

Lopresti's search for perfection reached its peak in the matter of safety. The slim Eppler wing would have to accommodate both a main landing gear and a fuel tank between its two main spars, located one third and two thirds of the way from the leading edge. And the attach points between wings and fuselage would be made far stronger than needed for aerodynamic loads, so that in case of impact the wings would stay on, absorb the crash energy and keep the cabin from rolling.

Finally there was the cabin itself. Federal regulations called for seats able to withstand 9 gravities of impact, as on airliners. Lopresti and his colleagues beefed up the cabin to far higher specifications. A strong metal keel would run forward under the engine. Under the entire cabin, with its floor of Space Age honeycomb panels, there would be 8 inches of crush space. The seats would all be designed for 25 gravities of impact rather than 9, and they would have several more inches of energy-absorbing crush space below them. The front seats, most vulnerable in the event of an accident, would have four-point safety belts like those Air Force pilots use.

Lopresti's obsession with safety on the 301 would cost weight— always the nemesis of aircraft designers. Almost 5 percent of the 301's empty weight would be accounted for by the extra strength, beyond aerodynamic requirements, needed to provide the crash-worthiness that Lopresti demanded. The problem solved itself by the very efficiency of Lopresti's design. His plane would fly so far on a gallon of fuel that its tanks need not be large. The pieces of the design had come together in one harmonious whole.

For the 301's potential buyers, most of them hard-driving and time-conscious, the ultimate sales argument would be how fast the plane would move through the sky, and how low a fuel bill it would run up. Lopresti checked every detail of the 301's design in the Cal Tech wind tunnel before the first sheet of metal for the prototype was cut. The measurements confirmed a top speed of over 300 mph for the Mooney M-301 at 25,000 feet. The 301, Lopresti's dream of excellence, is very close to reality. The prototype is already flying, and its performance confirms the wind-tunnel results. Mooney plans to market the big single around 1986.

❋

Like Roy Lopresti, Jim Griswold served a tour of duty with the Air Force at Wright-Patterson in Dayton. He too has also designed a whole new airplane starting with a clean sheet of paper. There is nothing out of the parts bin in his PA-46 Piper Malibu except the throttle lever.

Griswold began his professional career at Cessna in 1954, then in 1962 worked with Bill Lear as head of the testing department on the first Learjet, the Model 23. North American Rockwell lured him away to manage preliminary design for its Commander division, and at Rockwell he developed the pressurized Commanders, powered by Garrett 331 turboprop engines. He returned to Cessna in 1966 to manage Cessna's line of military aircraft, then its commercial aircraft. The Cessna Citation, easiest to fly of all the pure jets, was developed with substantial input from Griswold. He was brought to Piper in 1978, and soon afterward drew up the concept of the PA-46.

The plane was developed in considerable secrecy. Griswold's plane, like Lopresti's, is a fast, high-flying, pressurized six-passenger piston single. Its Continental engine is rated at 310 hp, but the engine's top-end components all come from a much larger Continental of 435 hp. The exhaust stacks of the big six-cylinder engine drive twin turbochargers, and the air they compress is chilled by two intercoolers before the intake manifolds. Careful tuning of both the intake and the exhaust gives an extraordinarily high fuel efficiency of .395 pounds per horsepower-hour. The wing, using the same airfoil as the Citation's, has an extreme 11:1 aspect ratio for high efficiency, and a span of 43 feet, larger than that of many twins. The PA-46 is designed to be flown hard in all weathers. Wing boots for deicing are driven by dual vacuum pumps, one of them clutch-driven and unused unless the first one fails. Dual alternators provide electrical power.

Griswold's design philosophy favors flight testing over wind tunnels, and a full-size aerodynamic mockup of the PA-46 was flown for 350 hours, including spin tests, before he began the detailed design. He argues that the customer buys the cabin, and that everything else serves to move that cabin quickly and safely. Where the M-301's cabin is large, the PA-46's is huge, and is pressurized to an 8,000-foot cabin altitude for flight at 25,000 feet. (That altitude, not an aerodynamic limit, is the highest that the FAA allows any plane to fly unless it has dual-pane windows.) The cabin interior of the PA-46 is large, but the fuselage exterior is not bulky, because wasted space has been cut to a minimum. Entry is through an airstair door behind the left wing. A baggage compartment in the pressure shell is supplemented by another big one forward, between the shell and the engine compartment. As a result, flight in the PA-46 is very quiet.

The first PA-46 Malibu off the Piper assembly line was delivered in 1983, and Piper (with the backing of its conglomerate parent, Bangor-Punta) committed several million dollars to a new factory for the plane, even in the midst of the worst depression the light-aircraft industry had known for decades.

Griswold's PA-46 and Lopresti's M-301 turn out to be remarkably similar aircraft, the first two of a new generation. The PA-46, marketed about three years before the M-301 and equipped initially with a slightly smaller engine, cruises at 245–255 mph. The M-301 can cruise at 270 mph, but burns 25 percent more fuel per hour than the PA-46 to do so. At the same fuel-burn rate, the two planes cruise at about the same speed.

The M-301 is a little smaller and more slippery aerodynamically, but the PA-46's efficiency-tuned engine provides somewhat more power at the same fuel economy. The PA-46's bigger tanks give it 15–30 percent more range than the 301, and it has more cabin and baggage space. But each of the planes can make a transcontinental trip with only one fuel stop in an easy day's flying.

For such aircraft, normally operated on a direct straight line between small airports convenient to the starting point and the destination, the distance that has to be flown is usually 25 percent less than the highway driving distance between the two points. So in terms of equivalent highway driving, the 301 owner commands a vehicle with an effective top speed of 400 mph. And throttled back to an economy setting of 16 gallons per hour, both the PA-46 and the M-301 deliver the highway equivalent of 340 mph average speed, while getting 21 highway miles per gallon—fuel economy comparable to that of a six-passenger automobile, and at six times highway speed.

With their pressurized cabins and six-place seating, their speed, efficiency and beauty, the PA-46 and M-301 may well become classics of design, rivaling such legendary planes as the DC-3, the

P-51 Mustang and the Beech Bonanza. If so, they will be flying well into the 21st century. For the present, they are executive transports only. But sometime within this next decade, the robotics revolution and the Geostar system should make them affordable and easy to fly. When that happens, many families who had never dreamed of owning an aircraft will have at their command a 340-mph "limousine."

In the marketing of transportation service, we can see a natural progression for the M-301, the PA-46 and others of their kind. In the mid-1980s, they will usually be flown by professional pilots with "Commercial" or "Air Transport Pilot" certificates. Most of the new planes will fly executives on company business. But when a family of four or five people want to make a vacation trip between towns of small or moderate size, up to a thousand miles apart, they will be able to charter one of the pressurized singles for a fast, comfortable, direct flight free of airline baggage hassles and interline connection problems. Their charter costs will be comparable to the cost of airline tickets for the same journey.

By the late 1980s, the Geostar navigational system should be in place. That will greatly simplify the piloting task, and the Geostar transponder linked to the most complete form of autopilot, called a "Flight Director," will make possible fully automatic landings in IFR conditions at most of the 15,000 airports in the United States. In that linkage, Geostar will give precise information on aircraft position, and the Flight Director will couple it with information on the plane's attitude in roll, pitch and yaw. Recent advances in microengineering have made possible new product lines of lightweight, relatively inexpensive, reliable Flight Directors. They are beginning to appear in single-engined aircraft.

One series of new Flight Directors is made by the S-Tec Corporation of Mineral Wells, Texas. They derive roll information from the simplest form of gyroscope, which is electrically driven and is held by springs so that it cannot tumble. Information on pitch is derived not from a gyro but from a very sensitive pressure sensor that measures minute changes in attitude. The heart of the sensor is a marvel of microengineering: a combination of an electrical capacitor and integrated circuit chips that is made for the Ford Motor Company as part of an auto-engine electronic fuel control. In its automated manufacture, a computer-guided laser beam trims the capacitor to high precision. The S-Tec Flight Director stabilizes the third axis, yaw, by sensing lateral oscillations of the tail with a simple accelerometer. For the smoothest possible automatic landings, the Geostar three-dimensional position information feeding a Flight Director may be supplemented by altitude information derived from an inexpensive ultrasonic altimeter. A sensing element precise to half an inch at altitudes below 30 feet costs only $150.

When the pressurized singles are equipped with the Geostar/ Flight Director combination, flying them will be so much easier that relatively low-time pilots with "Private" certificates can safely be allowed to rent them, either from plane-rental companies or from flying clubs. That will open a larger market for the planes, and will bring about price reductions by economies of scale. When aircraft rentals are available everywhere, as car rentals are now, it will become possible to rent a plane at one airport, fly it and leave it at another, at far lower cost than for a charter. Finally, we should begin to see a gradual phase-out of the present ground-based navigational system, to the point where new planes need not legally carry the thirteen different instruments. From then on, the only cockpit equipment required for safe flight will be dual redundant Geostar receivers linked to a Flight Director, at a total cost that is only a small fraction of the cost of today's cockpit electronics.

Market pressures will then be intense to go for the private-owner market, and therefore to automate the production of both aluminum and composite aircraft. For the latter, Brandt Goldsworthy's totally automated glass-fiber construction methods will come into play. Under the pressures of an expanding market, automation will spread to the small factories that supply subcomponents (engines, landing gear, electrical equipment, hydraulics) to the light-aircraft industry, and with that automation, prices will fall still further, opening a widening market.

As prices fall, more and more families and individuals will find it possible to invest in aircraft. Most of those planes will be equipped with Geostar/Flight Director controls, backed up by conventional manual controls. The minimal planes of that era, just as today, will be four-passenger, fixed-gear aircraft corresponding roughly to economy cars. The high end of the market, corresponding to luxury sedans like the Mercedes-450, will be filled by pressurized six-passenger singles like the M-301 and the PA-46. When the total inventory of aircraft reaches 20 million, with an average replacement interval of seven years, light aircraft will constitute an annual market of about $50 billion. That is consistent with present estimates by Airbus Industrie of the market for large commercial aircraft. Airbus projects that between now and the end of this century, the world's airlines will take delivery of 6,000 to 7,000 new aircraft, at a total price of $300 billion.

Just as intense road building began only when private cars were sold in large numbers, so the opening of the skies will bring about a new era in airport construction. That will reverse an unfortunate present-day trend, the disappearance of small airports. As our population has grown and spread from cities out to suburbia, housing developments have filled the open pastures around our older airports. Consequently, many of them have been hit by new noise-abatement operating restrictions. Harassed by lawsuits and tempted by soaring land prices, many airport operators have sold out to developers. At present, we are losing airports at the rate of about 300 per year. And federal specifications for the construction of new airports demand large, thick runways able to accommodate big aircraft. That has pushed the price of new airports up out of sight.

Fortunately, the loss of airports is a problem that individual states are beginning to solve. The government of Ohio, exasperated at the proliferation of federal airport regulations and conscious of the economic value of airports, inaugurated a program of its own to make sure that every county in Ohio had at least one good airport. The program was financed entirely by the state and its municipalities, without federal help. It was very successful. Michigan is following Ohio's example. A committee of the state legislature, recognizing "the public benefit of general aviation and its positive effect on Michigan's economic growth," recommended building at least one good airport in every county. Ninety percent of the total $265-million cost will be funded by the state, and 10 percent by the municipalities directly served.

I have made this detailed argument for large-scale growth in the private-aviation market because that potential growth is perceived by only a small number of people. Which nations will dominate the new market? Almost all the new designs are American, but both Europe and Japan will be formidable competitors as soon as they observe a growing market.

In the case of automobile production, the Japanese responded to a new market opportunity in the United States, brought about by new regulations and pressures for fuel economy, more quickly than did the Americans themselves. And while the market for light aircraft in Japan is unlikely ever to be significant, that will not deter the Japanese from seizing an opportunity for export. (As a parallel, the highest-power motorcycles manufactured by Kawasaki, Honda and Yamaha are made for export only, and are illegal on the roads of Japan.) Japan's very strong expertise in production engineering and in robotics, its abundance of highly trained engineers in all fields and the attractiveness of aircraft as value-added products all combine to suggest that the Japanese will be aggressive in penetrating the new market for aircraft designed for individuals and families.

The current history of large-aircraft construction is a disturbing guide to the possible future history of light-aircraft production. In the present stiff competition between the United States and Western Europe for the commercial-jetliner market, Japan is a kind of wild card. Both U.S. and European companies are negotiating to bring in a major Japanese company as a partner for the next round of airliner development. And each is fearful that the Japanese alliance may be made with the other. American antitrust laws would prevent the alliance of large American firms, like Boeing and Lockheed or McDonnell-Douglas, but they would not prevent a foreign alliance.

Japan is particularly interested in Boeing, which is widely regarded as the strongest company overall in design, manufacturing and marketing. Mitsubishi and other Japanese firms already make body panels, wing components and other parts for the Boeing-757 and 767 jetliners. One of Japan's strongest cards is the longevity of its companies. Many of the U.S. subcomponent manufacturers on which American plane makers once depended went under or abandoned the aviation market in the 1981–1983 depression. With Japan's more patient money and the help that MITI is already providing to its aircraft industry, Japan is quietly building the infrastructure of aircraft-component manufacturers that it will need for a complete, diversified aviation industry.

Construction plans for the Avtek 400 show what may be the entering Japanese wedge in the aircraft industry. Avtek is based in Camarillo, California, and its officers and consultants are mainly Americans. The Avtek 400's turboprop engines are PT6A-28s of 680 hp made by Pratt and Whitney of Canada. Dupont de Nemours, manufacturer of the Kevlar fibers and Nomex aramid honeycomb that provide most of the plane's strength, is closely involved with Avtek, as is the Dow Chemical Company. But the advanced, all-composite airframe is to be manufactured entirely in the Kobe-Osaka-Shiga area of Japan, by Avtek Far East, whose President is Dr. Kigen Kawai. Dr. Kawai was one of the first developers of carbon-fiber composites, and will benefit from consultation with Toray Industries in Japan. Brandt Goldsworthy calls Toray "the most advanced company in the world for commercial production of exotic fiber composites." As of mid-1983, Avtek had escrowed $100,000 from each of 88 prospective buyers of the Avtek 400. The selling price of the plane is about $1.4 million, which will make it highly competitive with other business aircraft in its class.

Quite possibly, if present trends continue, Japan may become the only nation outside the U.S.S.R. with the total capability for civil-aircraft manufacture. The United States will have much to regret if that point is reached just when the new market for light aircraft— potentially far larger than the market for commercial aircraft—is ready to explode.

CHAPTER 7

*E*L DORADO IN ORBIT

For the first twenty-six years of the Space Age, until well into 1983, all the dollars earned by space hardware came from the performance of unmanned satellites launched into orbit by unmanned rockets that were used only once. But the liquid- and solid-fuel rocket-engine technologies that made those launches possible were developed over nearly half a century, primarily in Germany, the Soviet Union and the United States, for a mix of manned, unmanned, military, commercial and scientific programs.

UNMANNED LAUNCH VEHICLES

The reliability of unmanned launch vehicles has increased steadily through those years. The liquid-fueled Titan III rocket has been the workhorse for U.S. Air Force satellite launches for more than two decades, and the basic rocket engines of the Titan III, built by the Aerojet Liquid Rocket Company, have worked without a failure in 130 consecutive launches spanning seventeen years. More than 160 nonmilitary satellites have been launched on Delta rockets, made by the McDonnell-Douglas Astronautics Company in Huntington Beach, California, and the Delta's record has also been very good. The Delta began operations in 1960, and in its first fourteen years demonstrated a reliability of 91 percent. Since 1974, following substantial improvements, Delta rockets have been successful in 96 percent of their launches. Delta's long history can be used as a guide to likely future developments in unmanned vehicles for commercial satellite launches.

Evolution of the Delta has been so dramatic that almost nothing remains of the original design. By 1982, there had been twelve successive Delta models. The Delta of 1960 could put only 75 pounds into the geostationary orbit where most communications satellites must work. The Delta of 1982, a far larger vehicle, placed 2,050-pound satellites in geostationary orbit, and still further upratings to the 2,800-pound range are planned.

All the improvements made during the Delta's long history have been based on a conservative design philosophy: in every upgrade, only flight-proved, space-qualified hardware is used.*

In 1980, as the Space Shuttle era was about to begin, McDonnell-Douglas developed a new third stage, the Payload Assist Module (PAM), which attaches to a satellite payload rather like an outboard motor to a boat. It can be mounted either on top of a Delta rocket or in the payload bay of the Shuttle. The PAM carries solid propellant and is stabilized by spin. It furnishes the thrust to lift a payload satellite from low Earth orbit to geostationary orbit. Late in 1982, millions of people watched on television as a PAM carried in the Shuttle payload bay came up to full rotation on its spin table and then ejected from the Shuttle before firing at a safe distance.

With the rapid buildup of the commercial satellite business, the backlog of orders for Delta launches has grown significantly. In 1982, it was necessary to reactivate Launch Complex 17B at NASA's Kennedy Space Center on Cape Canaveral, to achieve a capability of 10 Delta launches per year. Present bookings for the Delta go well into 1986.

Delta's history shows that there is no real separation between military and nonmilitary or between manned and unmanned systems in the development of rocketry. Much of the present engineering effort focuses on the development of more completely controllable engines and more easily storable fuels. Sophisticated new missiles like the MX require final-stage engines capable of being throttled over a wide range from low thrust to maximum, and shut down and restarted frequently, in order to follow complicated evasive maneuvers directed by onboard computers. An MX fourth stage developed by Rocketdyne has a main (axial-thrust) engine that must be restarted 21 times during a simulated fifteen-minute mis-

---

* The current Delta uses as its first-stage engine the Rocketdyne RS-27, which burns kerosene (RP-1) and liquid oxygen. RS-27 engines were left over at the conclusion of the Apollo program. The Delta's second-stage engine is derived from an Air Force Titan upper-stage engine, as modified by work done in supplying a second-stage engine for a Japanese N-II rocket.

sion. It also has eight attitude-control engines that must shut down and restart some 2,600 times during the same period.

TRW has derived from the Apollo Lunar Descent Module engine another missile-control engine which can be throttled over a 10-to-1 thrust range, and can be run for thrust periods from steady state to bursts as short as eight milliseconds. That technology can be returned to peaceful application in commercial ventures when large-scale construction projects are carried out in space, requiring frequent docking and precise maneuvering of orbiting spacecraft.

Even when a satellite is in orbit, it still needs rocket motors. Its position ("station") must be kept within small limits to offset the forces of lunar and solar gravity, further gravitational forces caused by the Earth's not being a perfect sphere and the pressure of sunlight on its solar-electric panels. Its attitude (orientation) must be held correct, against all those forces and against torques caused by the Earth's magnetic field, in order that its antennas may be oriented correctly toward points on the Earth.

For many years, the tasks of station keeping and attitude control have been carried out by small thrusters using a single propellant, hydrazine. That chemical can be stored for orbital lifetimes of as much as seven years, but when passed through a catalytic converter it decomposes, with the release of energy. More energy can be obtained from the burning of a fuel with an oxidizer, but until recently no good oxidizer with a long storage lifetime was available. The Bell Aerospace Textron company then found a way to produce highly purified nitrogen tetroxide, and its lifetime was long enough that it could be loaded into the Ford Aerospace Insat 1A satellite. In 1982, a Marquardt engine burning storable fuel with nitrogen tetroxide in the Insat 1A ran for two burns of about half an hour each, to place the satellite in a perfect circular geostationary orbit. The engine will continue to fire as necessary to provide station keeping for the satellite during its lifetime.

JAPANESE ROCKETRY

Outside the United States, the Soviets, the Japanese and the West Europeans have comparably high levels of technological expertise in rocketry. India and the People's Republic of China have successfully launched satellites, but the rocket programs in those nations have not yet matured sufficiently for us to assess their long-term potential.

The Space Activities Commission of Japan is made up of five

men, oversees Japan's several space programs and reports directly to the Prime Minister. Japan's policy on space development, established more than twenty years ago, calls for fully public (unclassified) work and stresses international cooperation. Japanese activity in space is substantial and mature, with a growing budget that is currently about one tenth of NASA's. (If one compares the Japanese space budget with NASA's unmanned, non-Shuttle activities, the U.S. and Japanese programs are more nearly equal.)

There are two satellite-launching programs in Japan, both administered by the Commission. The Institute for Space and Aeronautical Sciences (ISAS), originally a part of the University of Tokyo and now independent, has developed the "M" series of solid multi-stage rockets for launching scientific satellites into elliptical, relatively low orbits. Beginning with the launch of the 50-pound "Osumi" test satellite in 1970, the ISAS program has progressed at the steady rate of about one launch per year, from the ISAS Kagoshima Space Center at the southern tip of Kyushu. The ISAS M3-S rockets, built by Nissan Motors, currently put satellites of 440-pound weight into elliptical low orbits.

The National Space Development Agency of Japan (NASDA) was established in 1969, primarily for developing larger, liquid-fueled rockets to launch satellites for communications and weather observation. Since the beginning of Japan's space research in 1960, there has been a very close and cordial relationship between the U.S. and Japanese space programs, and the United States licensed much of the Delta rocket technology to Japan to assist in the start-up of NASDA.

Over the past few years, Japanese policy has been directed toward building up independent launch capability. NASDA launches are carried out on the island of Tanegashima from a launch complex about 70 miles south of Kagoshima. The first NASDA launch, in 1975, placed a 180-pound satellite in orbit with the N-I-1 rocket, built by Mitsubishi. In 1977, a later N-series rocket placed a 290-pound satellite in geostationary orbit, and by now N-II launches, occurring about once per year, place medium-size satellites of 750 pounds in geostationary orbit.

For still larger satellites, NASDA is developing the H-I rocket series, with a first stage using the N-series (Delta-like) kerosene/liquid-oxygen engine, and an entirely new high-performance second-stage engine, the LE-5, that is fueled by liquid hydrogen and liquid oxygen. That LE-5 "cryogenic engine," with about 50 percent more thrust than the main engine of the comparable U.S. Centaur rocket, was first tested in 1982. It ran for about twenty minutes (regarded by rocketeers as a substantial period of time), on its test stand, and did so at constant thrust and pressure with no instabilities or ignition problems. With the high thrust and long burning time of the LE-5 engine, the H-I rocket, built by Mitsubishi Heavy Industries, should begin its growth career in the mid-1980s by launching 1,200-pound satellites to geostationary orbit.

Drawing on Japan's electronics and instrumentation industries, the nation's space program has broad diversification, and its satellites, built by firms like Mitsubishi Electric and Nippon Electric (NEC), are of high quality and performance. The Mitsubishi ETS-1 satellite of 1981 was 100 percent Japanese, and the H-IA vehicle will have about 90 percent Japanese components (by hardware cost), contrasting with 35 to 56 percent in the earlier N-series rockets. Japan's long-range program calls for the launch of multiton space platforms by the latter part of the 1980s, and for the possible commencement of independent manned space-flight operations sometime after 1995.

Japan has so far given no indication of entering the market for commercial launches, but if it does so, its H-series rockets will be formidable competition.*

EUROPEAN ROCKETRY

Western Europe has entered the commercial launch market in a big way, with its Ariane rocket. The Ariane is a version of the Diamant, a rocket developed independently by France in the 1960s. Just as Japan's cryogenic LE-5 engine gives the H-series rockets high performance, the Ariane achieves a heavy lifting capability because of its own new cyrogenic engine—developed, like the LE-5, without detailed technology transfer from the U.S. space program.

All Arianes lift off from a pad within the French space complex at Kourou in French Guiana, just north of Brazil. Kourou is only 5 degrees in latitude from the Equator, which is a double advantage for the launch of geostationary satellites. The Earth's rotation of

---

* Commercial (nongovernmental) satellites are presently built for voice and digital data communications, and to relay television programs.

1,000 miles per hour at the Equator gives a free boost of almost that speed from Kourou for launches eastward, and the eventual plane change necessary to match the satellite orbit to the equatorial plane costs little in rocket fuel when the difference in planes is only 5 degrees. Cape Canaveral, at 28 degrees latitude, and Tanegashima, at 30 degrees, are at a comparative disadvantage for launches to geostationary orbit.

At the beginning of the 1970s, the U.S. Government had decided to abandon the development of its expendable launch vehicles, which were then, technologically, five to ten years ahead of all competition. Instead, the United States concentrated on the reusable Space Shuttle. But repeated delays in Shuttle development, resulting from inadequate funding, provided an excellent opportunity for other nations to catch up. By 1980, France was able to persuade its partners among the European nations to form Arianespace, a private company based in France, to exploit the growing market for satellite launches.

There are fifty shareholders in Arianespace, and the largest is the French space agency, with 34 percent of total ownership. Another 25 percent is held by French companies. West German firms hold 20 percent of the shares, and the remaining 21 percent are spread among nine European countries, all of them involved in Ariane hardware production. Ariane's shareholders include thirteen European banks, and most of the remaining shareholders are companies building Ariane components.

At a Satellite Summit Conference held in Washington late in 1982, Frédéric d'Allest, the President of Arianespace, was challenged on the subsidies given his firm by European governments. He retorted that those subsidies amounted to no more than 1 percent of the $1.6 billion subsidy given by the U.S. Government for the first three years of Shuttle operations. Arianespace is pursuing an aggressive financing policy, something that is becoming, increasingly, a feature of commercial satellite operations worldwide. A customer need put up only 15 percent of the launch cost at the time an Ariane rocket places his satellite in orbit. He has five years to pay off the remainder, at low interest rates (11 percent as of 1983) provided by the member banks among Arianespace shareholders.

Because Arianespace has never enjoyed the high level of governmental funding that developed America's rocket technology, the firm has been forced to commit itself to commercial operations relatively soon in its career. The results have been mixed. Three of the Ariane's four test flights were successful, but its first commercial flight, late in 1982, failed, dumping the ESA (European Space Agency) satellites Marecs-B and Sirio 2 into the Atlantic Ocean. The company then emphasized that Flights 5 through 9 were really "transitional." Despite the failure of Flight 5, ESA and other prospective customers remain in line for future launches. Ariane Flight 6, in mid-1983, was fully successful.

The U.S.-built Westar 6, GSTAR 1 and 2, and Spacenet 1 and 2 satellites are all to be launched by Arianespace. The company had firm bookings for the launch of thirty-two satellites from Kourou as of early 1983. It is working toward a rate of six launches per year from one pad, and in 1985 it will bring a second launch pad into operation at Kourou.

On Flight 6, Arianespace began experiments on the recovery by parachute of the rocket's first stage for possible reuse. The first version of the Ariane has a lift capability very close to that of the Delta rocket: 6,000 pounds to low orbit, and 2,100 pounds to geostationary orbit. The uprated Ariane 44L, with four strap-on liquid rocket boosters for extra thrust, will be operational in 1984, boosting 2-ton payloads to geostationary orbit. The heavier satellites which the Ariane 44L can lift will be able to carry more propellant for station keeping, important because satellites normally run out of propellant before their electronics quit working. An added year of satellite life is worth typically $20 million in revenues.

PRIVATE COMPANIES

So far there are just two private (totally unsubsidized) ventures that have come close to orbital capability. The first, OTRAG, is based in West Germany and began its launch tests from sites in equatorial Africa. But its host nations were politically unstable, and its operations were subjected to an intense propaganda barrage from the Soviet Union. Those political problems have so far kept OTRAG from commercial success. Another company, Space Services, Inc., of America, hopes to provide usable commercial launch services by the mid-1980s.

The American firm is the creation of David Hannah, a Texan educated in civil engineering and business, who built the company as a second career after thirty years as a land developer. Hannah's long-term goal is to develop, construct and launch satellites for commercial clients. His first customers may be oil companies who need to survey the Earth from low orbit, with color-sensitive cameras that can pick out the clays characteristic of oil-bearing formations.

Hannah hired a small California firm to design and build a rocket for low-orbital launches, to be called the "Percheron." There was nothing wrong fundamentally with its design, but the first Percheron blew up on its launch pad in 1981 (no rare event for the first trial of any new rocket), at a cost to Hannah of more than $1 million. His second try, made with the good advice of NASA retirees Max Faget, a design leader in the Apollo and Shuttle programs, and astronaut Donald (Deke) Slayton, was based on buying government-developed hardware. The group chose the well-tested but obsolete Minuteman, a three-stage solid-fuel rocket. Throughout the 1970s, a small private company, Space Vectors, Inc., had been checking out surplus Minutemen in Utah. After checkout, the rockets were launched from the White Sands Missile Range in New Mexico, to carry sensor payloads into the upper atmosphere for the Air Force. Hannah's major challenges in the following year were political, but he was operating in a favorable environment. The Reagan Administration looked kindly on private ventures, NASA wanted to get out of the commercial launch business and there were nearly fifty surplus Minutemen, under Department of Defense jurisdiction, at NASA's Wallops Island research facility on the Virginia coast.

Hannah accumulated $6 million in financing. In October of 1982 he launched the "Conestoga," a second-stage Minuteman carrying a dummy payload of a half ton of water, into a 300-mile trajectory from Matagorda Island, off the Texas coast. Hannah wanted to begin low-orbital launches of 500-pound payloads by 1984, using all three Minuteman stages. However, the Department of Defense refused to turn over more Minuteman rockets to Hannah's firm. One of his alternatives is to keep open the Convair–General Dynamics assembly line for the Atlas-Centaur rocket and to persuade lending institutions to regard Atlas-Centaurs as bankable assets for eventual launches.

The Atlas, the kerosene/liquid-oxygen rocket, is a well-tested first-stage vehicle, and many Atlases were built in the early days of the missile program. For years they were available as surplus, and NASA launched fifty-six Atlas-Centaurs in succession, all carrying peaceful scientific payloads, without a single failure. (Some other Atlas-Centaurs, kept stored in missile silos, failed when tested.) The Atlas-Centaur's lift capability is comparable to that of a Delta or Ariane—about a ton to geostationary orbit—so Hannah's acquisition of the rights to buy Atlas-Centaurs and launch them would give him the full-service capability he is looking for.

Among all the vehicles used to launch commercial satellites, the Space Shuttle towers above the rest in size and sophistication. It completed its series of test flights and began launching commercial satellites only in 1983, so there are not enough data yet to judge how it will stack up in competition. It has no competitors for those rare payloads consisting of extremely large satellites or "antenna farms" serving many users from a single site. But most of the launch traffic is in satellites of moderate size, and there the Shuttle's advantages are not so clear.

As the development histories of the Delta and the Ariane show, the throwaway rockets can be uprated and reconfigured, even changed in their gross outlines, because aerodynamics does not matter to them. They go through the atmosphere only once, and that is at a relatively low speed just after lift-off. It is routine by now to design such throwaway rockets with "hammerhead" payload fairings much larger in diameter than the rockets themselves, to accommodate wide satellites. Similarly, it is no great matter to lengthen a payload fairing to enclose an especially long satellite. By contrast, the Shuttle must enclose payloads in its bay, because the Shuttle is a complete aerodynamic vehicle that must reenter the atmosphere at a speed of 18,000 mph.

With the Shuttle carrying two satellites in its payload bay, each attached to a PAM-D solid rocket for the subsequent lift to geostationary orbit, Burt Edelson, NASA's Associate Administrator for Space Science and Application, quotes a (subsidized) launch price of $18,000 per pound of satellite. The price for a Shuttle launch is set by U.S. Government policy rather than by profit-and-loss in the sense of business accountancy, and under existing policy, that $18,000 per pound will go to about $31,000 per pound for launches not yet contracted. The Ariane 44L should be able, according to D'Allest, to launch to geostationary orbit for about $28,000 per pound.

The Ariane's designers argue that the Shuttle is volume-limited rather than weight-limited as far as commercial satellites are concerned. But if we consider weight alone, the Shuttle will be able to launch satellites of half-ton weight for approximately $16,000 per pound by around 1986. It will do so for eight satellites at a time, in a manner that has long been argued over: by carrying in its payload bay a Centaur rocket, fully fueled with liquid hydrogen and liquid oxygen. Once in low orbit, the Shuttle will open its payload-bay doors and release the Centaur, carrying a framework with the eight satellites mounted in it. In a series of maneuvers involving multiple restarts of the Centaur's engine, the rocket will then climb out to the geostationary orbit and put each satellite at the longitude that its task requires.

There are three chief disadvantages to this planned mode of operation: the obvious risks of putting a rocketful of explosive liquid hydrogen and oxygen in the Shuttle payload bay for the ascent; the penalty in weight (now thought to be 6,000 pounds out of the Centaur's total 14,000-pound payload) for the hardware framework necessary to tie eight satellites, each of 1,000 pounds weight, to the Centaur; and finally, important for commercial competitiveness, the relatively inflexible scheduling that must be imposed when eight different satellite-construction programs must all converge to the same launch date. But some scheme of that kind appears to be the only way that the Shuttle can achieve a substantial price advantage over its competition.

The Shuttle is not very well matched to the task of emplacing average-size (half-ton to one-ton) geosynchronous satellites. The vehicle will be better employed in missions where the destination is low Earth orbit and its download capability is used. But that is a problem for NASA rather than for the commercial satellite user. To the user, a launch vehicle is simply a means to an end: the emplacement of a payload. The near-term business opportunities in space all depend on what such payloads can do. To judge, it helps to review the developments of the past two decades. As Andrea Caruso, Secretary General of the European regional network Eutelsat, says, "A decade in technology is like a century in politics."

SATELLITE USE

From 1965 to 1982 the commercial-satellite market grew nearly 100-fold—a compounded annual growth rate of 30 percent per year. In the same period, satellite costs dropped to one fifth of their 1965 prices. The first network, Intelsat, was developed on a philosophy of intergovernmental cooperation, and its goal was to provide international telephone service competitive with undersea cables. One hundred six nations, of all political persuasions, are members of Intelsat, and about $2 billion is now invested in that network. Comsat, a quasiprivate corporation, is the U.S. member of Intelsat. It is also the U.S. member of Marisat, which provides specialized communications services for ships at sea.

As the markets for satellite services have grown and costs have decreased, many new entities, both governmental and private, have entered the arena. With their entries, the issues of profits, competition and regulation have become controversial. Article 14 of the Intelsat agreement of 1970 specifies that "no significant harm" to Intelsat is to be permitted from newer satellite networks. Nations other than the United States, eager to exploit new opportunities, argue that Article 14 should not apply to market opportunities that Intelsat would like to exploit but has not yet invested in. Intelsat, clearly, would prefer instead to preserve its semimonopolistic position even for markets it hasn't yet had time to penetrate. But it seems that Intelsat is fighting a losing battle because of the large economic incentives that tempt its competitors. The attractions of a satellite system from a business viewpoint are rapid market penetration and high return on investment. As Clay Whitehead, President of Hughes Communications,* said, "With satellites, a very small company with a specialized market niche can have access to the entire U.S. market on Day One."

The revenues can be substantial, though small compared with those of any top-30 corporation. Edelson of NASA estimates that the total investment in satellite systems in the past twenty years has been $8 billion, and that current revenues are $3 billion per year. A nation or a company can rent a single circuit on a satellite, capable of supporting one voice-quality telephone conversation, for about $390 per month—down from $2,600 per month in 1965. That is about a penny per minute, and the ultimate user is paying about 200 times that rate when telephoning to foreign countries.

A satellite, despite its cost (typically $30 to $100 million), often

---

* Whitehead left Hughes in 1983 to start a new company whose goal is to finance and operate communications satellites.

pays back its full price in its first year of operation. But the risks are also high. Satellite Business Systems, a computer-data networking system that now has 16 corporate clients and about 100 "Earth stations" with the capability to send signals up to the SBS satellites as well as to receive from them, is a joint venture of IBM, Comsat and Aetna Insurance. Yet SBS may have hit the market a little too early, because its Comsat investors alone are in the red by $300 million. Robert Hall, President of SBS, notes that the satellite links themselves are very reliable, but that all his problems result from the thousands of land lines that are needed to connect the Earth stations to the ultimate users. Those have far higher failure rates than the satellite links, and are labor-intensive for error detection and maintenance. The same problem plagues Intelsat. Though the satellite link itself costs the user only half a cent on every dollar he spends on an overseas phone call, the connecting land lines that complete the call are so much more costly that they absorb most of the rest. Andrew Inglis, who retired in 1983 as President of RCA American Communications (RCA Americom), notes that RCA became profitable in satellite communications only in 1979, after fifteen years of investment—and that as soon as it did so, some members of the Federal Communications Commission wanted to regulate it, to limit profits. Today's $3-billion-per-year income for the industry looks good, but Inglis believes that total losses over time still exceed total profits in satellite communications.

For communication between fixed points on Earth, satellites must compete with other technologies, such as fiber optics. MCI is a lower-cost competitor to the Bell System (MCI is tagged "Money Coming In" by those who count its revenues). MCI is installing a fiber-optics cable from New York to Washington, D.C., rather than using a satellite link between those points. Andrea Caruso argues that satellite links become cost-competitive only between points more than 500 miles apart. But for communication between points where population densities are low and no infrastructure of land lines exists, satellites pay off. And as the number of independent channels (transponders) on each satellite has grown, now to as many as 48, each with a bandwidth capable of carrying 900 two-way telephone circuits, it has become practical for developing nations and new companies to lease transponders while waiting for their own satellites. Eutelsat leases unused transponders on the French Telecom-1; the American Satellite Company leases 20 percent of the transponder capacity of the Westar satellites, while preparing for the launches of its own two satellites in 1985 and 1986. (It has 128 Earth stations and serves major corporations, banks and the Federal Government with high-capacity links for data and for tele-conferencing. The Gannett daily newspaper *USA Today* depends on American Satellite for the high-speed facsimile transmission that permits Gannett to print the same newspaper in a number of cities simultaneously.)

Satellite communication will be constrained increasingly by the limited range of frequencies (bandwidth) available to it. Fiber-optics communication, however, will have no such limit, because extra fiber bundles can always be added to an existing line without inter-ference. Already new satellite services are being forced into com-paratively unfavorable frequency ranges because the older ones are saturated.

Within the microwave range, frequencies of 2 to 3 gigahertz (2 to 3 billion cycles per second) are attenuated least by clouds, rain or hail. But those frequencies, in what is called the "S band," were used as early as World War II, and the S band was occupied by terrestrial point-to-point microwave links and other users before the first communications satellite was launched. Satellite services were therefore allocated frequencies in the C band: near 4 gigahertz (GHz) for the satellite-to-Earth-station downlink, and near 6 GHz for the uplink. In the C band, attenuation by water in its various forms is not very bad, though worse than in the S band. But com-panies introducing new services are now being forced into the K-u band (12 GHz and 14 GHz for the two directions) or the K-a band (15 GHz and 17 GHz). At those frequencies, atmospheric water attenuates the transmissions much more severely. The SBS com-pany is using the K-u band, and finds that on clear days only 1 bit out of 100 million that it relays is garbled. But when the signal path from an SBS satellite to one of its Earth stations goes through rain, the error rate increases 100-fold. For the few minutes of passage of an intense desert rainstorm, the error rate may be as bad as 1 in 10,000, and then the system has to be shut down.

While a substantial fraction of the available bandwidth is used for the interchange of computer data, the most voracious users of sat-ellite communications are cable networks. Television programs are sent up to a satellite which relays them to Earth stations at cable "headers." From each header a network of cables takes the programming to the customers. Cable companies can select different programs on the same downlink frequencies if their Earth-station receiving antennas are large enough to be highly directional; they can select a signal from a particular satellite just a few degrees in orbit away from its neighbors. As of 1982 there were already 5,000 different cable-TV companies in the United States, serving about 30 million customers. One supplier of Earth stations alone, Scientific Atlanta, had by then built more than 6,000 receiving antennas for cable headers. But there is so far no way to prevent customers from pirating the downlink signals, and Scientific Atlanta's president, Sidney Topol, estimates that there are more than 40,000 (illegal) backyard C-band antennas.

It was volume and the Gallium Arsenide Field Effect Transistor ("GasFet") that brought the cost of Earth stations down low enough for local cable networks to proliferate. The GasFet is a solid-state device that can amplify signals in the C and K bands, while adding very little noise. With still greater-volume production of GasFets, more elaborate signal-processing techniques and satellite transmitters of 200 watts power (about 5 times as much as in existing transmitters), it will be possible by 1986 to receive television programs at individual homes directly from the new satellites, with receiving antennas only 3 feet in diameter. Irving Goldstein, Executive Vice-President of Comsat, estimates that there is a potential market of an additional 30 million homes in the United States, not now served by cable networks, for programming from such direct-broadcast satellites ("DBS"). Eight companies, including Comsat, have filed applications with the Federal Communications Commission to provide DBS services.*

NAVSTAR

The market for DBS can be penetrated in other ways—for example, by lower-cost cable systems. In that application, satellite services will be competing for market share. But while nearly every conceivable satellite service has a competing earthly counterpart, for some applications satellites have such overwhelming advantages

---

* There is plenty of money to be made by selling DBS receivers at $1,000 or $2,000 each. It is not so clear, however, how to bill the customers for the programs being sent down. There may well be a flourishing trade in "blue boxes" that unscramble whatever encryption the DBS companies put on their signals.

that their market share is likely to approach 100 percent in the long run.

The most critical of those applications is to meet the needs of aircraft: for precise navigation information, for collision avoidance and for precision approaches to airport runways in bad weather. There are currently two satellite systems, one partially completed and one in an early stage of development, that can satisfy some or all of those needs.

The first is GPS, the Global Positioning System, developed by the U.S. Department of Defense during the 1970s and 1980s. It employs the Navstar satellites. GPS was developed so that military aircraft could navigate anywhere in the world without revealing their positions. It is a relatively complex system, including in its fully developed form 24 satellites in orbits with a twelve-hour period, 21,000 km above the Earth (about halfway out to geostationary altitude). Those satellites orbit in three different planes, so that some of them are always in view from anywhere over the globe, including the poles. Each satellite carries an atomic clock that can maintain accuracy within a thousandth of a microsecond. The satellites broadcast complex, encrypted signals at two different frequencies. Receivers on board the aircraft decode the signals, compare them in order to correct for varying delays in transmission time introduced by the ionosphere, and finally present the data to the pilot.

The GPS/Navstar system has had a somewhat checkered career. Its primary problems are cost and complexity. Only about half the Navstar satellites are in place, so there are currently substantial gaps in coverage, depending on location and the time of day. Already an atomic clock on one of the Navstars has failed, so that one satellite is broadcasting information of reduced accuracy, based on timing from its backup quartz clock (essentially the same technology as in a wristwatch). Because the cost of the Navstar system has escalated greatly since it was first proposed, present plans call for only 18 of the 24 satellites to be orbited. That leaves very little margin for redundancy, so the failure of one or more additional Navstars would leave the users with more gaps in coverage.

High officials of the DOD estimate that the final cost of the Navstar system will exceed $30 billion. But the same arguments that led to the design of GPS/Navstar have also convinced the Russian military that it must follow the same route. In 1982, the

U.S.S.R. announced that it will orbit a system of "Glonass" satellites (Global Navigation System) forming what is really a clone of Navstar. Glonass will use the same technology and nearly the same frequencies.

The high costs of their military-oriented systems have led both the U.S. Congress and the Soviet Government to propose that civil users buy receivers for them and pay user fees, to help finance the two projects. At the same time, the DOD (at least prior to the announcement of Glonass) was reluctant to release to civil users the full information (the so-called Precision Code, or "P Code") that would permit getting high-accuracy locations from Navstar. The DOD feared that the Soviet Union would obtain the code and use it for missile attacks on U.S. missile silos.

The response from civil users to the DOD proposal has been less than enthusiastic. The U.S. airlines are disinclined to sign up for expensive services. The airlines of the whole world, who coordinate their technical work through ICAO, the International Civil Aviation Organization of the U.N., have formally rejected by letter the American DOD proposal, because they do not want to become dependent on an expensive military system whose services might be denied them in any extended period of international tension. And the users of light aircraft have been wary of a system that would require still another expensive piece of onboard equipment. Of the three major needs of aircraft for location information, Navstar can provide only one, en route navigation—unless the DOD releases the P Code. With the P Code, civil aircraft could make precision approaches using Navstar, as the DOD intends military aircraft to do. But because Navstar is a one-way information-transmission system, returning no information about positions to any central monitoring station, it is incapable of protecting aircraft against midair collisions.

GEOSTAR

The applications of the developmental Geostar Satellite System to aviation were discussed earlier. Ironically, the wholly peaceful Geostar system does not carry the risk that Navstar does of providing military information to an enemy. While the GPS/Navstar system is vulnerable to penetration of the P Code, in Geostar no useful information comes back to the user until it has been processed through Geostar's central computer on the ground. A hostile aircraft or missile could obtain information only by transmitting the fingerprint of a registered Geostar transponder—and the computer would instantly recognize that the fingerprint was not coming from the last airport where the transponder's aircraft landed.

Although Geostar is designed to cost less than 1 percent of the price of Navstar, aviation users alone will be too small a market to support even the less expensive system. Fortunately, Geostar will have many other markets, because of two unique features: its ability to transmit digital information in two directions, between satellites and small, inexpensive (even hand-held) transponders; and its channeling of all information through a single complex of computers on the ground.

With an inexpensive hand-held, battery-operated transponder about the size of a pocket calculator, any backpacker in the wilderness can call for emergency help instantly—and have the nearest rescue service notified within half a second of his exact location. A woman threatened by an attacker on a dark street at night can push a button and know that within half a second the Geostar computer will notify the police vehicle nearest to her, and will tell its officers her exact location and identity. Further, within another half second she can have positive confirmation through a light or sound from her Geostar transponder that the nearest police vehicle has received her message. When a car is stolen, Geostar can notify the nearest police vehicle instantly of the point of theft, and can add from the memory banks in its central computer the make, model, color and license number of the stolen vehicle.

In the developing nations particularly, communications services are frail and break down soon after the start of any natural catastrophe. As the Geostar transponders will be inexpensive, rescue vehicles even in the Third World can be provided with them, and the Geostar computer can be used as a central clearinghouse for information, relaying it to the nearest Geostar satellite for transmission to those in charge of the rescue operation. To provide the entire world with such services, eight Geostar satellites will be needed. All will be in the same (geostationary) orbit at different longitudes, so two more satellites can be added as in-orbit spares. Any one satellite can take over the functions of another simply by moving to a new location in geostationary orbit. Such changes require very little energy.

Motorists can be guided to their destinations in towns unfamiliar to them, without risk of getting lost. In case a car breaks down, no matter how remote the location, Geostar can locate the nearest available tow truck and pass on to the stranded motorist a confirmation within seconds that it is on its way, then guide the tow truck to the location of the breakdown. Boaters can navigate by Geostar to stay within safe channels, even at night or in fog. Fishermen can return to favorite spots even miles at sea, and find them within errors of a foot or two.

But Geostar's ability to send messages both ways doesn't end with the exchange of information between the central computer and one particular transponder. The dispatch office of a long-haul truck fleet or a delivery-vehicle fleet can send to individual drivers information on new pickups to be made, and receive prompt information from drivers on any problems. Geostar's computer will add the vehicle's identification and location to the message before sending it via satellite to the dispatch office. The capabilities of Geostar are so great, in fact, as to raise a concern about privacy. Yet individuals who subscribe to its service can always turn off their transceivers at any time.

Future geostationary satellites like Geostar, and more than 150 geostationary satellites already orbiting, show how space can be used effectively and profitably. In the building and launching of those satellites, there has been no need for dramatic breakthroughs in aerospace technology. But the boundaries of that art have been extended greatly during the past decade through the long, patient and often thankless work of NASA and its contractors on the Space Shuttle.

THE SPACE SHUTTLE

The *Columbia*'s first orbital flight on April 12, 1981, and the string of successful flights that followed it, proved how well that work had been done. The Shuttle extends launch-vehicle technology in two ways: it is largely reusable, and it can return bulky, multiton payloads from orbit almost as gently as a commercial airplane lands people and cargo. The Shuttle has been written about so extensively that I will only touch on its basic numbers. We will learn more by exploring its subtleties and especially its long-term potential.

At lift-off, the Shuttle assembly weighs 2,250 tons—about 25 percent less than the old Apollo/Saturn 5. Its two solid-rocket boosters, built by Thiokol in Utah, each generate 1,450 tons of thrust. Its three main engines, built by Rockwell International's Rocketdyne Division and running at 3,000 pounds per square inch of pressure, burn liquid hydrogen and liquid oxygen more efficiently than any rocket engines ever developed before them, to add 188 tons of thrust each at sea level. Three minutes later, in vacuum, their thrust is 25 percent higher. All five engines are gimbaled so that their thrust directions can be altered by signals from flight computers. The delta-winged Orbiter vehicle itself carries no fuel on board for its main engines. Instead, those draw from the largest single component of the Shuttle assembly, the external tank. That tank, 154 feet long and 27.5 feet in diameter, carries 780 tons of the cryogenic propellants, and the engines drain it dry in less than eight minutes at nominal full power. Although the tank, weighing 37 tons empty, stays with the Orbiter almost until orbital velocity, on all present and foreseeable NASA flights that tonnage then reenters the atmosphere and burns up.

After the external tank is released, the final nudge to low Earth orbit comes from two engines built by the Aerojet Liquid Rocket Company, each engine providing 3 tons of thrust. Their fuel is monomethyl hydrazine, and it ignites spontaneously on contact with its oxidizer, nitrogen tetroxide. (Such a mixture is called "hypergolic." It is used where absolute reliability is essential and where the propellants must be stored for a long time at room temperature. In the Apollo program, it was used for the ascent stage that lifted each pair of astronauts safely from the Moon to orbit.) Those two engines also serve for all subsequent orbit changes: to alter the orbital plane or altitude, to rendezvous with a satellite or space station, and finally, as retrorockets, to brake the Orbiter's speed enough to cause it to reenter the atmosphere. In space, the Orbiter needs more thrusters to permit small excursions in all three directions as well as rotations in roll, pitch and yaw. There are 44 of those smaller thrusters, and they burn the same hypergolic mixture as the 3-ton-thrust maneuvering engines. Thirty-eight of the rotational thrusters are of 870 pounds thrust, while 6 are "vernier" thrusters of only 25 pounds.

All cargo is carried in the Orbiter's payload bay, 15 feet in diameter and 60 feet long. The weight that can be lifted to orbit in that bay depends sharply on the orbital height and plane: 65,000 pounds to a very low orbit 100 nautical miles above the Earth, in launches due east from Cape Canaveral; only half that much for polar launches from Vandenberg Air Force Base, just south of Santa Barbara, California, and only 25,000 pounds for launches into a 55-degree plane at an orbital height of 270 miles.

But the Shuttle was not designed to be a weight hauler. Its original purpose was to provide routine transportation both ways between the Earth and a large manned space station in low orbit, and for that it would excel. It can carry up to seven people, and maintain them in comfort in a normal sea-level atmosphere. Its payload bay is big enough to accommodate large subassemblies of a complete low-orbital station. For maximum crew safety, it has redundancy for all its systems, like a commercial aircraft. And finally, it is intended for quick turnaround. Rockwell International, its prime contractor, had to design it so that each Orbiter could land, go through all its servicing and refurbishment, be mated with its external tank and booster rockets, and be ready to launch again on two hours' notice within a turnaround time of just two workweeks of five days each, with two eight-hour shifts per day.

The Shuttle represents a real breakthrough in aerospace engineering not in its launch capabilities, but in the Orbiter's reentry. It makes a transition over a half-hour period from orbiting at 18,000 mph to landing on a runway as a very fast but otherwise conventional aircraft. Its touchdown speed, 190 knots, is not drastically different from the Concorde's approach speed of 162 knots. The remarkable achievement of its design is that it remains under control, aircraftlike, throughout that speed range of nearly 100 to 1, while its energy of motion is gradually bled off into heat radiated from its thermal-tile skin.

The Orbiter's designers achieved that control by abandoning a once-hallowed tenet of aeronautical design: natural aerodynamic stability. Left to itself, the Orbiter would tumble. Stability is forced on it by its flight computers, which control not only its elevons but a "body flap" at its tail and a speed brake under its fuselage. Thanks to those extra surfaces and the programs carried in its five redundant computers, the Orbiter handles very much like an ordinary airplane. Although the Shuttle astronauts have hand-flown the Orbiters to perfect landings a number of times, on many Shuttle missions the approach and landing are flown by the onboard computers, acting as the world's most sophisticated Flight Director.

The Orbiter is very much like a conventional airplane. Aluminum ribs, stringers and skin form its structure. Its landing gear, working by a simple gravity drop assisted by the airstream, looks much like the gear of a medium-size jetliner. Only its outer skin of tiles, not really a structural part of the Orbiter, reveals that it is meant to scream through the atmosphere at 30 times the speed of sound.

With the first Shuttle orbital flight, it became apparent that its designers, many of them veterans of the Apollo program, had done their work exceedingly well. It also became apparent that each one of them had "fudged" a bit in the direction of safety. Reentry temperatures were lower than the 2,300 degrees Fahrenheit that the wing leading edges were designed for; the acoustic noise level in the payload bay at lift-off was lower than expected; peak structural loads were lower than target values. The Orbiter's main landing gear was designed to sustain hard braking, but after the first few flights, the astronaut crews were letting the plane roll out to a smooth stop with no braking at all.

For all the success of the Shuttle's maiden voyages into space, *Columbia* and her sister ships were put into service for missions very different from those for which they were designed. It may be some years before they are used appropriately enough that their unique capabilities can really be exploited. In the absence of a U.S. space station, they are having to perform double duty as both space station and the transport to it, and they cannot do that without compromise.

Designed as commuter vessels for quick forays into orbit followed by immediate return to Earth, they have two-week turnaround capability—which now is being wasted because they must wait in orbit for seven days (and by the mid-1980s, as long as thirty days). Designed on a mission model of 120 flights per year, they are now reduced under budgetary pressures to a schedule of only a few flights in the early part of this decade, increasing to a modest 24 flights per year toward 1990. NASA now estimates that there will have been only 300 Shuttle flights in all, even as late as 1992. With less use, the price to users will rise to $90 million per flight for flights scheduled after 1985.

Some Shuttle trips will use its manned capabilities effectively: those on which it carries the European-built Spacelab experimental station up to orbit and back. On others, like the planned launch of the Orbiting Astronomical Observatory, the Shuttle's main advantage may be that the payload will not be lost in a launch failure. The Shuttle's people-carrying ability will hardly be used on missions in which it lifts to orbit four Delta-class satellites, together with their PAM rockets, for the journey to geostationary orbit; or lifts one heavy satellite together with a fueled-up Centaur rocket. (The Centaurs that will ride in the Shuttle payload bay are identical only in their engines and controls to those which have traditionally carried satellites in Atlas-Centaur launches. The Shuttle versions will be short, fat machines only 29 feet long, with fuel tanks 14 feet in diameter to fill the Shuttle's payload-bay diameter.)

NASA's commitment for more than a decade to the successful development of the Shuttle left the agency with neither the funds nor the attention that was needed to fill the Shuttle's payload bay with valuable cargo. That role was left to Europe and Japan. Eleven European nations, acting through ESA, have developed the Spacelab, which will ride up to orbit, remain several days and return to Earth, all without leaving the payload bay. In orbit, as many as five specialists will drift through a tunnel to the Spacelab from the Orbiter's two-level crew compartment, and in a shirt-sleeves environment will control experiments remotely mounted on an attached pallet filling the remainder of the payload bay. Those experiments, some purely scientific and some directed toward small-scale industries in space, will all depend on the zero-gravity environment, the high vacuum surrounding the Shuttle, the Earth's wrapping of magnetic fields or the unobstructed view of the universe that a low-orbital site permits.

West German firms are interested in casting high-performance turbine blades in the zero-gravity environment because it is free of convection, which comes from weight differences. Professor T. Obayashi of the University of Tokyo has designed for Spacelab-1 a comprehensive experiment on plasmas (clouds of ionized gas) in magnetic fields. It will have an accelerator to project electron beams, an arc jet to produce an intense argon plasma and a nitrogen jet to project neutral gas into nearby space. With those tools he expects to produce and study artificial auroras, excite and study plasma waves and explore couplings of plasmas to the Earth's magnetic field that are suspected of connecting the Earth's weather to activity on the Sun. While the Spacelab's design and development have been carried out by ESA, the United States is expected to purchase a Spacelab of its own at some future date.

CIVILIAN VS. MILITARY USE

The original charter of NASA called for it to be entirely separate in its missions and its operations from those of the Department of Defense. Even as late as 1969, the Air Force planned for its own wholly independent manned space program, whose first goal was to be the launching of a manned orbiting laboratory. Budgetary constraints then forced the cancellation of one of the two manned space programs, and the Administration chose to cut off the military program, forcing an uneasy marriage between NASA and the Air Force.

Then, as the microengineering and robotics revolutions advanced during the 1970s, the Air Force became less and less convinced that there was any military need for men in space at all. But with its own launch vehicles scheduled for decommissioning under further budget cuts, it had no choice but to accept the Shuttle to carry its payloads of reconnaissance cameras, radio-emission detectors, heavy radars and other information-gathering systems. Much has been written about the "militarization" of the Shuttle, but the ironic fact is that the Shuttle was forced on the Air Force rather than taken over by it.

In the long run, the Shuttle will probably take its place with other vehicles that are sold to different customers for different purposes. The KC-97 jet tanker was developed by Boeing for the Air Force, then turned into the Boeing-707 for the civilian market. The Electra turboprop was developed as an airliner, but many of those planes were also sold to the U.S. Navy to carry submarine-detection equipment. So far there is no evidence that the peaceful NASA space program is being compromised by the assignment of some Shuttle flights to the Air Force. Some missions will be all-civilian; others, flown by Air Force personnel, will be all-military; and there will be no mixing of the two.

As for how the Air Force will use the Shuttle, the best clue is the location chosen for its own Shuttle launch site and landing strip: Vandenberg Air Force Base. Vandenberg is ideally suited to the launch of reconnaissance satellites into polar orbits from which, over a twenty-four-hour period, they can map and survey the entire Earth for signs of military activity.

NASA estimates that it still has several years of work to do before the Shuttle's minor defects are cured. At the same time, Jim Beggs, NASA Administrator, is arranging to have the turnaround servicing done by a single company, for efficiency, and to model that servicing on the routine maintenance characteristic of an airline, rather than on the constant tinkering that goes with test flights. When Shuttle operations do become routine, and not only the marketing but the operation of the Shuttle is put into private hands, as Beggs intends, it will be time to consider Shuttle upgrades. Like all aerospace vehicles, the Shuttle has growth potential. A recent study by F. L. Williams of Martin Marietta Aerospace explored the easiest, least risky ways in which the Shuttle could be improved, and in which specialized vehicles could be derived from it.

Williams first tackled the problem of lifting unmanned payloads one way, from Earth to low orbit. For such tasks, the airplanelike portion of the Orbiter's 150,000-pound dry weight—58 percent of the total—is nearly useless, and there is no real need for the Shuttle to carry a crew. Williams designed a simple freight vehicle that uses the Shuttle's booster rockets, external tanks and Orbiter main engines just as they are now. The Orbiter airframe is replaced by a large, simple payload shroud, with the main engines located and angled as in the conventional Shuttle.

The vehicle, unmanned, could lift a 150,000-pound payload to a 160-nautical-mile circular orbit in a launch due east from Cape Canaveral. That is 2.6 times the payload capability of the present Shuttle to the same altitude. The main engine and controls either would be recovered as a package, equipped with parachutes and automatic braking rockets for a soft landing, or would travel back to Earth in the payload bay of a subsequent Shuttle flight that would otherwise return empty. Fortunately, the 64,000-pound weight of the main engines, controls and supporting structure is just within the regular Shuttle's 65,000-pound return-weight limit.

Williams' more ambitious plan, one that has been considered many times by NASA in various forms, is to replace the solid rocket boosters by liquid rocket boosters of higher thrust, burning cheaper fuels. For minimum development cost, Williams chose to drive each booster by five of the standard Shuttle-Orbiter main engines, burning either their regular mixture of hydrogen and oxygen, or else a storable hydrocarbon fuel with oxygen. Because the Orbiter's capability would then be limited by the volume of its payload bay rather than by weight, that replacement would not be of any particular value on regular Shuttle flights. But on freight-only, unmanned flights with the large expendable payload shroud, the Shuttle-derived freight rocket could lift a payload of 185,000–200,000 pounds—about 3.5 times the present Shuttle's payload to the same altitude. Flights in that configuration would actually be cheaper than those of the unmodified Shuttle.

Another evolution of the Shuttle, requiring substantially more development but offering still lower cost per flight, would be to give the liquid rocket boosters small wings with which they could fly back to an airplanelike landing on the Shuttle runway. In that form, the booster design could follow the well-developed technology of cruise missiles. In the long run, those developments could roughly halve the cost per pound for dense payloads carried within the Orbiter's payload bay, and could reduce to one fourth the launch cost per pound for cargo lifted one-way to orbit.

THE SOVIETS AND THE SHUTTLE

All of these possible upgrades to the Shuttle would cost far less than did the present Shuttle system. If they do come about, the rationale could be either economic, with the start of substantial exploitation of the energy and material resources of space, or nationalistic, as we observe the progress of the U.S.S.R. toward the establishment of a permanent orbital colony.

So far, the Russians do not have a Shuttle of their own. Throughout the 1970s and early 1980s, they stayed with tried-and-true expendable launch vehicles. Their response to the U.S. development of the Shuttle was twofold: an unrelenting propaganda campaign against it and the quiet development of their own alternative. Soviet diplomats always refer to the Shuttle as a "weapon," and in August of 1981, the Soviet delegation to the U.N. introduced a draft resolution, citing the Shuttle by name, that would have banned all "weapons" from space. (The Russian war of words is accompanied by more dangerous preparations; U.S. intelligence reports leaked to the magazine *Aviation Week* indicated that the 15-ton vehicle *Cosmos 1267*, launched in June 1981, was armed with missiles capable of destroying space vehicles, including the Shuttle.)

The Russians' own competitor to the Shuttle is far along in development, but it is following a route that the United States decided —rightly or wrongly—to bypass. The concept is that of the "Dyna-Soar," a small, winged glider launched on an expendable booster and capable of returning crews to the Earth. Dyna-Soar models were tested successfully in a series of suborbital flights beginning in

1963, but the concept was abandoned in the United States with the demise of the Air Force's independent manned program.

The Dyna-Soar makes a great deal of sense in many ways. To lift a payload to orbit takes as much energy as lifting it up a mountain 4,000 miles high. Once it is there, economy suggests leaving it there and augmenting it with other payloads, rather than returning it to the Earth only to be hauled up again at great expense, as we are doing with the Spacelab.

The Russians reason that there is just one component of an upward-bound payload which always has to be brought back: the crew. With a small glider capable of reentering the atmosphere and landing, they can recover that crew as cheaply as possible. That leaves them the freedom to launch freight on either manned or unmanned flights, and to send a crew along, in the recoverable glider, only when that crew has essential business in space. Separating the functions of crew recovery and payload lift in that way, they have much more freedom than our Shuttle gives us to lift payloads of awkward dimensions or odd weights. In June of 1982, the Soviets launched a one-third-scale model of their recoverable reentry glider in an unmanned flight from their Kasputin Yar launch site.

It appears that while the United States has concentrated on launch-vehicle development, the Soviet Union has chosen to concentrate on technology for the permanent occupancy of space. The Russians have perfected the technique of docking, which we brought to a fine art in the Apollo program but then left unused, and they have cautiously extended to more than half a year the tours of duty for crews in orbit. Their target, based on routine practice in Arctic and Antarctic research stations, is to exchange crews once per year.

They have also made progress in recycling systems for life support. On *Salyut 6*, a mechanical system working on the humidity in the spacecraft recycled 50 percent of the water. The next step is a urine still, which could lead to 90 percent recovery. They have tested banks of *Chlorella* algae for the recycling of carbon dioxide. That could lead to 80 percent recovery of oxygen. Food plants have been grown in the Soviet spacecraft, at first as a novelty but now as a significant dietary supplement, particularly for vitamins. Wheat, cucumbers and cabbage are favorites in the Soviet program. By the mid-1980s, some 20 to 40 percent of the bulk food eaten by the cosmonauts in orbit will be grown there. One puzzle remains: while plants can be grown from seeds in the zero gravity of orbit, they do not flower and produce a second generation of seeds. As the radiation level in orbit is insufficient to prevent flowering, the cause of the problem is almost certainly the absence of gravity—a problem that later, rotating habitats will avoid.

While space stations of substantial size can be assembled in orbit from components no larger than the Soviets' present launchers can lift, the U.S.S.R. is also developing larger rockets. Following the death in 1967 of Sergey Korolyov, the inspirational leader of the Russian space program, the Russians failed dismally in their development of a big booster in the Saturn 5 class. After leaving that attempt abandoned for more than a decade, they have now, apparently, resumed it. Persistent intelligence reports and published articles from writers with contacts in Eastern Europe all indicate that the Soviet Union is developing a booster somewhat larger and less efficient than the Saturn 5, but with about the same payload capabilities: 200,000 pounds to low Earth orbit, and roughly 100,000 pounds to geostationary, to lunar orbit or to escape into interplanetary space.

The old launch pads for such giants are being refurbished. The present best guess of the Soviet-watchers is that by about 1985 the U.S.S.R. will use one of those rockets to launch a 100-ton space station for permanent occupancy by crews of twelve to twenty cosmonauts. Newspaper and magazine interviews with leaders of the Soviet space program give a consistent picture of Russian activities in space for the remainder of the century: gradually assembling a colony in Earth orbit, "Kosmograd," with several hundred inhabitants; towing a large station to lunar orbit as a base for repeated excursions by cosmonauts to the Moon; and possibly using that same technology for a two-to-three-year manned voyage to Mars. Some reports indicate that the U.S.S.R. has already assembled three hundred of its brightest children in Moscow and is training them at a special school for space colonists at the Palace of Pioneers. Their future role is planned: to live and work in the orbiting Kosmograd by about 1995.

THE ECONOMICS OF THE SHUTTLE
While the Soviet Union is pursuing a logical long-term plan for gradual expansion into space, a plan driven by national policy

rather than by economics, the West is concentrating more and more on deriving economic benefits from space. There is at least one substantial commercial possibility in space that has never been discussed, and it would match the Shuttle's unique capabilities better than any program currently planned. The one task for which the Shuttle has no rival is the safe, gentle recovery of heavy payloads from orbit. And there is one valuable payload that always requires round-trip service: people.

A great many people, young and old, would willingly save for years to be able to make one unforgettable journey into space. The travel time is short enough—less than half an hour either way—that they could put up with spartan accommodations en route. Once in orbit with some 325 paying passengers, the Shuttle could rendezvous with a comfortable space station, rotating to provide artificial gravity in most of its public rooms. And there, for two to four weeks, the visitors could observe the lands of the Earth below them; visit observatories for spectacular views of the universe; attend lectures on history, geography and astronomy; enjoy the sensations of zero gravity when they choose and dine well on food brought up on their own Shuttle flight. At first hearing, the idea sounds frivolous, but if realized, it would pay its own way and open a unique opportunity for many people who have a burning desire to experience flight in space. And if the reports of the astronauts and cosmonauts are any guide, those people would come back with a heightened sense of global community that could only benefit the world.

What is now called "Space Industrialization" is a precursor to bringing two-way space flight into the economic picture for the first time. It means the production of very special products: those which can be made only in zero gravity, and whose value per pound on Earth is so high as to justify lifting their raw materials into orbit. Some pharmaceuticals sell for much more than the Shuttle-transport cost of roughly $1,400 per pound, so pharmaceutical companies are cautiously investing in pilot plants to make in orbit those few of them which can benefit from zero gravity. The market for products of that kind is so specialized that it can never grow very large. But there is another kind of space industry with a high potential for growth. To understand it, we need to consider the relation between market volume and product cost.

The first phase of commercial utilization of space has been based on the transfer of information, which weighs nothing. To go beyond that phase requires selling material products at prices competitive with alternative products made on Earth. But all the material products on which large terrestrial industries are built are relatively inexpensive in cost per pound of weight. The costs per pound of cereal grains or gasoline are as little as 10 to 20 cents. For commercial-grade metals, they are typically 50 cents to $2. Heavy appliances and automobiles cost from $1 to $5 per pound, and electronic products roughly $10 to $150 per pound. Almost no mass-produced goods can be sold for much more than that. But transport costs to orbit and back would add $3,000 per pound to the cost of any product made in space from Earth materials. Therefore, no large-scale industry can grow on the basis of raw material lifted from the Earth, processed in space and returned to the Earth for sale. To escape that limit, we need to find products whose point of sale is not here on Earth, where they must compete with goods being sold at $1 to $150 per pound, but in space itself.

SOLAR-POWER SATELLITES

The first phase of commerce in space, now successful and mature, is based on one such product: satellites for the gathering and the relay of information. There is at least one other material product whose point of use would be in high orbit: solar-power satellites (SPS).

The SPS concept in its original form is due to Dr. Peter Glaser of the Arthur D. Little Company in Cambridge, Massachusetts. His idea was audacious but logical: to put in geostationary orbit, where there is permanent sunshine, an array that converts solar energy to radio energy; then to send that radio energy in a beam to a receiving antenna on the Earth for conversion to ordinary electricity. The SPS concept has been studied over the past fifteen years by NASA, the Department of Energy and the National Research Council. The key numbers, reviewed many times, appear to make sense. Radio waves could be produced efficiently and at high power anywhere in the microwave S band or below. Those waves could be directed to a receiving antenna in the form of a flat array several miles in extent, held on poles above fields and meadows. Sunlight and rain would penetrate the array, but the microwaves would not, so land below the antenna could be farmed or grazed. The microwaves could be converted to electricity with an efficiency of more than 90 percent—far higher than the efficiency of a coal or nuclear power plant. Environmental effects have been studied with particular care, and the SPS appears to be quite benign, with no measurable effect even on insects and birds that fly through the densest part of the radio beam.

The potential market for power satellites could be very great indeed, if SPS plants could be sold at a cost competitive with that of coal or nuclear plants. A typical SPS would supply 10,000 megawatts (10 gigawatts, abbreviated GW) to the power lines on Earth. Its salable value delivered to geostationary orbit would be at least $10 billion, because it could supply that power without using any fuel. The installed electric-generator capacity worldwide as of 1980 was about 10,000 GW, and the average life of a generating plant was about thirty years. Replacement alone with SPS plants would therefore open a market with a potential of about $300 billion per year worldwide. Growth at 3 percent per year would create another $300-billion annual market. Penetration of those markets even to the 10-percent level over the next decades would open a new industry with annual revenues of $60 billion; substantial penetration could open an industry comparable to the worldwide automobile industry.

The problem with that rosy picture is transport costs. It would take fairly exotic engineering, with correspondingly high costs, to reduce the weight of a 10-GW SPS to 100,000 tons. For it to sell competitively with coal and nuclear, the price of the power station in geostationary orbit would have to be well under $50 per pound. But transport costs from the Earth, at Shuttle rates, would be at least 60 times as great. (Lifting SPS components from low orbit to geostationary would roughly double their transport cost from the Earth.) To reduce the transport costs to 30 percent of the final sale price would require bringing those costs down to a two-hundredth of the present level. An independent committee of the National Research Council, set up to review the SPS concept, refused to believe the transport costs could be brought that low.

It seems that the problem of transport costs could be overcome by a radically different approach which builds on two of the growth markets we have already covered: microengineering and robotics. It also builds on a fact known since the time of Newton: that the energy required to lift an object from the Moon to a high orbit is less than a twentieth as much as to lift it from the Earth. We could build satellite power stations out of lunar materials, and take advantage of the Moon's low gravity and vacuum environment to transport those materials to orbit at low cost. We could accelerate materials at the lunar surface to a speed so great that they would escape the Moon entirely.

The most efficient machine for that purpose is called a "mass driver," and it is a form of linear synchronous electric motor, based on the same physics that was used in designing the Transrapid-06. The mass driver would be shaped as a pipe, 16 inches in diameter and 800 feet long. The pipe would consist of a stack of aluminum coils. Within the hollow pipe another coil, the "bucket," would move. It would be supported and guided by magnetic fields, and would hold in a cavity at its center a baseball-sized payload of sintered lunar soil. The bucket would be accelerated through the pipe by magnetic forces produced by currents discharged successively into the outer "drive" coils. Those currents would be taken from a solar-cell array nearby.

Computer-aided design has led to a plan for the mass driver. It specifies a total weight of 10 tons for the complete installation, including the solar-cell array. The first two thirds of the machine would provide the acceleration for the loaded bucket, to a speed of 1.5 miles per second. The last one third would decelerate the empty bucket to a stop. The acceleration cycle would be repeated every five seconds, and with operation during the 35 percent of the time that the Sun is high in the lunar sky, the mass driver would transport annually some 800 tons of lunar material—80 times its own weight—to a point high above the Moon. The payloads would climb out of the Moon's gravity, losing speed each second, until finally after sixty hours they arrived at the collection point.

One other novel device is required if lunar materials are to be used for construction: a chemical-processing plant to separate the lunar soils into pure elements. Thanks to the Apollo project, we know very well what those elements are. The lunar surface consists of minerals very similar to clays found on Earth, but without their water content. By weight, it is about 30 percent metals, 20 percent silicon and 40 percent oxygen. Studies of the processing plant indicate that it could separate about 100 times its own weight in lunar soil each year. The technologies for the mass driver and the chemical-separation plant for lunar soil are now being developed through funding from a unique institution, the Space Studies Institute (SSI).

SSI, based in Princeton, New Jersey, is unique in that it supports developmental research entirely through contributions from individual citizens. It takes no government money and receives only a small fraction of its support from industry. SSI's mass-driver research is carried out through a grant by SSI to Princeton University. Three successful working models of mass driver have been built. The latest, with a cross section and design acceleration equal to those of a final lunar machine, agrees closely in performance with the computer program by which it was designed. Research aimed toward a processing plant for lunar soils is being done through an SSI grant to Rockwell International, at the same Downey, California, facility where many Shuttle designers work. In the first two years of that research, completed in 1983, Rockwell specialists under the direction of Dr. Robert Waldron measured all the chemical reactions necessary for the separation that were not already well known from previous industrial chemistry.

Experts in several fields assembled in workshops sponsored by SSI to find the most cost-effective way in which a space industry could be established, based on the mass driver and on the chemical separation of lunar soils. The participants operated on the rule "Invent as little as possible." They took the Space Shuttle in its present form as the only Earth-to-orbit transport system that would be available. They assumed a vehicle for transport between orbits of the kind that is in fact already being developed out of the Centaur launch vehicle. They went back to the Apollo Lunar Module to see how a reusable lunar lander could be derived from the Centaur. They took as given the technologies for life support and communications that have already been tested in space. And they took from industrial experience the observation that it saves development costs to parallel identical machines for each increase in production volume rather than to scale up to a larger machine.

In the SSI plan that emerged from the workshops, there would be just three units to develop that had no close counterparts in today's space hardware: the mass driver, the chemical-processing plant and a "job shop," a general-purpose complex of machines able to shape and weld metal. Each of the three units would be small enough and light enough to constitute a single Shuttle payload. The first two would be designed for automatic, unattended operation. The job shop, while highly automated, would be monitored and controlled through radio and television from the Earth.

Once the three units were fully checked out, both on the Earth and later in low Earth orbit, they would be emplaced on the Moon, and copies of the processing plant and job shop would be put into a high orbit between the Earth and the Moon. Then, controlled from the Earth, they would go to work. The processor and shop on the Moon would turn out the simplest, heaviest, most repetitive components of a second mass driver: the drive coils and solar-cell arrays. The principal by-product of the lunar processor would be oxygen, which would be liquefied and used as propellant for the lander. (The lander's Centaur engine would burn a mixture that was 15 percent hydrogen and 85 percent oxygen by weight.)

The processor and job shop in orbit would also turn out the simplest, heaviest components of a second processor and job shop. An additional Shuttle flight would deliver to low Earth orbit the light, complex, labor-intensive components needed to complete a second, identical set of units on the Moon and in space: computers, communications electronics, automated controllers, precision machine components. Finally the second set of units would be complete, and with it the production capacity would be doubled.

About eight doublings, estimated to take less than three years in all, would be enough, without any development of new hardware, to reach a production throughput of a quarter million tons per year. That capacity, turned to the fabrication of SPS plants, could turn out one per year, with a salable value of about $10 billion per year. As the materials for those SPS plants would come almost entirely from the Moon at low cost, the plants would be simple, rugged— and correspondingly heavy. Their primary energy conversion would probably use either amorphous silicon cells or turbogenerators.

The SSI plan is proceeding by ordered priorities on a five-year schedule that should see its completion about 1987; but as its research continues, there will be design work on other essential items. A new kind of SPS must be designed, under the ground rules that its weight does not matter, but that almost everything in it must be fabricable from lunar materials. With the completion of SSI's plan, any company, nation or consortium that has sufficient technical expertise will be able to implement it. Implementation is estimated to take five to seven years, including the three years of capacity doublings that will construct the mature space industry. Both that time and the estimated investment required are comparable to those of the Alaskan Pipeline, a wholly private venture.

Opening a substantial industry in space, based on materials from the Moon and on the constant, reliable solar energy found in high orbit, is by far the most speculative commercial opportunity of those we have examined. But it is also the biggest opportunity of all. It could free Earth's industries from their terrestrial limits and also solve one of humanity's greatest problems: how to supply clean energy for long-term worldwide economic growth. Remarkably, every study so far carried out shows that it could become profitable within a decade. Manufacturing in space carries such implications for national as well as commercial rivalry that we cannot yet tell what nation or company will be the first to mine the Moon and build power satellites; but within six months after it begins, there will be others imitating it.

THE FRONT-RUNNERS

Of the six opportunities for major new markets, microengineering is dominated by the United States and Japan. Technology for robotics was developed mainly in the United States, but Japan has become the world leader in selling robots. Europe runs a poor third in both those areas. Original research in genetic engineering is led by America, with Europe a strong second. Economically, genetic engineering is rich in long-term promise, but unlikely, for a variety of reasons, to become a major new growth market within the next decade. When it does so, Japan, already leading the world in fermentation technology, is likely to dominate production.

Among the last three opportunities, none of them yet developed to the market stage, West Germany is clearly in the lead to exploit magnetic flight for high-speed surface transportation. Japan has a strong program based on the more advanced technology of superconductivity—a technology not yet employed outside research laboratories. The United States, after more than a decade of governmental inaction on magnetic flight, has no developmental program at all. Any hope for America to compete in magnetic-flight systems is dependent on some highly innovative nongovernmental financing scheme, and on skipping past the Japanese and West German designs to the ultimate form of magnetic flight: travel in vacuum. Fortunately, that development would not require any advance in magnetics beyond the technology that the West Germans are using at this time. Whatever the system that eventually dominates, the worldwide market has been shown by West German studies to be several hundred billion dollars.

In private aircraft, the United States has two advantages: some of the managers of its light-aircraft industry are beginning to think in terms of automation, and the Geostar system for computerized flight has a reasonable chance of acceptance as a nongovernmental investment. In Japan, private aviation is virtually unknown, and in Europe it is regulated almost out of existence by socialist governments that view it as a "rich man's privilege." The market potential for light-aircraft sales within the United States is at least $50 billion, and the worldwide market should become hundreds of billions annually when the U.S. example is imitated.

In the construction of solar-power satellites in space, the game remains open. The United States struck out on its first try, because the main supporters of the satellite power concept here insisted on an unworkable technical approach—building huge tonnages of components on Earth, then rocketing those components to high orbit against the pull of gravity. The only practical research toward solar-power satellites is being followed through private support at a modest level in the United States. When that research is published, Japan, Western Europe or the United States could implement the results in the form of an action program. The Soviet Union is also interested in satellite power, but it may lack the ability to build the highly automated lunar mass drivers, chemical processors and fabrication machinery on which a cost-effective production program could be based. The first payback from space construction could be in the early 1990s, and the potential worldwide market is very large —several hundred billions of dollars per year.

Five of the six technology areas are now or could become within the next decade major growth markets. Of those, microengineering, the most mature, is a close race between the United States and Japan. The Japanese are ahead in robotics. West Germany leads in magnetic flight. The United States is well positioned to exploit the light-aircraft and space-construction markets, but will have to move swiftly or lose its lead in both arenas.

THREE

WHERE THE UNITED STATES LEADS

WINNING AMERICAN STYLE

In the last years of the 20th century, America needs to increase its productivity fast enough to regain competitive leadership in world markets. The new wealth would maintain a nonnuclear armed force strong enough to discourage the outbreak of World War III, build new industries that would provide full employment through the next decade, and help to educate our citizens for productive and satisfying lives in a new century that will be upon us by the time today's first-graders graduate from college.

Wealth-generating increases in productivity have been missed by American industry during the past fifteen years. In Part I we examined that failure, and the reasons for Japanese success during the same years. In Part II we explored major growth markets, open to several countries, which America must enter in order to generate new wealth. We can enter those markets only by finding a way to build new systems that are socially highly beneficial but not too novel or long-term to attract traditional private investment. Meeting that challenge may require a new financial invention, just as the settlement of our New World four centuries ago required the invention of the "corporation."

In Part III, we will see how Americans are developing new companies and increasing productivity in long-established firms. Here we will deal with American success. First, we will investigate an extraordinarily successful, uniquely American industrial/financial institution: venture capitalism. Then in the final chapter we will see how a few very large American companies have been able not only to survive but to grow in spite of the fiercest competition possible from domestic and foreign corporations.

CHAPTER 8

THE RISK TAKERS

Bookstores on the San Francisco Peninsula sell a board game called "In the Chips." Its subtitle, "Silicon Valley—the Local Investment Game of the Santa Clara Valley," overlies an artist's panoramic view northwest as it might be seen from the cockpit of a 727 on approach to San Jose Municipal Airport's Runway Three Zero Left. The landmarks are there: the elephantine hangars at the Navy's Moffett Field, built for dirigibles long gone; Stanford University's Hoover Tower and the dish of a radio telescope, among the live oaks on the brown foothills. Far up the Bay, a jet takes off from the international airport, and beyond it on the horizon are the high-rise towers of San Francisco and the long line of the Bay Bridge. The picture's viewpoint underscores a reality: San Francisco is beautiful and cosmopolitan, but the money in the Bay Area is being made elsewhere, well down the Peninsula in the heartland of Silicon Valley.

Wealth is created in Silicon Valley because of the combination there of five essentials: a great university (Stanford) which encourages the formation nearby of new high-technology companies that draw knowledge and new ideas from its faculty; a desirable residential area with pleasant weather and suburban lifestyle; easy access to an exciting city, San Francisco; a pool of professionals, skilled workers and would-be entrepreneurs educated and experienced in high-tech science and engineering; and several dozen venture-capital groups.

Venture-capital funding is a uniquely American invention. It means the investment of modest sums of money in new, generally high-technology companies with the knowledge that the investments carry high risk, but may yield high gains. Most of the successful venture capitalists have undergraduate degrees in science or engineering, and either M.B.A. degrees or practical operating experience in the management of technology-based companies. Each venture capitalist operates nearly independently of his colleagues, though generally two or more venture capitalists share a corporate identity.

The classical format of a venture fund is a partnership made up of two or more venture capitalists, who serve as the General Partners and make all the decisions for the firm. Their fund of money, usually at least $10 million and rarely more than $100 million, is put up by ten to fifteen Limited Partners, typically insurance companies and pension funds. The General Partners review proposals made to them by entrepreneurs, generally at the start-up or early phase of a new company. A small fraction of all proposals receive funding, of typically less than $1 million. Usually within the first year of buildup of the new company, one or more further investments are put in by the original investors. Once all the money in a fund is invested, the fund runs for seven to ten years and is then "cashed out," with the flow of accumulated value going to the General and Limited Partners.

Venture capitalists expect that most of their investments will be lost, but that a few lucky hits with very successful start-up companies will much more than pay for the failures. A well-run venture-capital fund may show a value increase of 35 percent per year compounded over its lifetime: a $200-million final value in ten years from initial investments of $10 million. The financial attraction of venture-capital investment is that it can return higher capital appreciation than any other form of investment in our society. It is also, clearly, not for the fainthearted.

Venture capitalists generally invest in manufacturing rather than service companies. Some of the best-known start-ups that received early funding from venture capitalists are Apple Computer, the ROLM Corporation, Tandem Computer, Federal Express and MCI. Of those, the last two are unusual both in their scale ($25 to $80 million in funding) and in being service rather than manufacturing companies.

The style of operation of venture capitalists is so individual that one cannot give a standard prescription for it. Instead, we will examine the individuals themselves. To begin, we will see venture capitalism from the viewpoint of someone who deals every day with the capitalists and the entrepreneurs. He is Mario Rosati, a lawyer.

Rosati went through U.C.L.A. as a major in Greek and Roman history. Mario couldn't have made a greater transition from the classical world. On meeting him, one senses energy, enthusiasm and a kind of boyish eagerness. He is dark-haired, slimmish and still only in his mid-30s. After receiving his Bachelor's degree, Rosati went to Berkeley's Boldt Hall, the top law school of the California university system, and sought out Richard Jennings, an expert in corporate law. He ranked number one in Jennings' course, but to the surprise of both of them, didn't get offers from Los Angeles and San Francisco law firms upon his graduation in 1971. Jennings advised Rosati to go down the Peninsula, where "a lot of things in electronics" were happening.

A law firm that had been started by John Wilson, Roger Moser and Pete McCloskey (who later, and for many years, was a member of Congress representing the mid-Peninsula) hired Rosati. Soon after that, he met Roland Jang, whom he describes as his "corporate mentor." In 1971, Jang was smarting when a company he had served for many years had been acquired by a larger firm. In the acquisition the two founders had made millions, but Roland and the rest of the staff had received nothing. He had vowed that in companies he founded on his own, there would be "something for everybody."

He proposed to sell Rosati $200 worth of founders' stock in a new start-up firm called International Medical Technology (IMT), which planned to make intensifying screens to yield high-quality X-ray pictures with lower dosages to patients. Mario bought 2,000 shares of IMT at 10 cents a share. Eighteen months later, Jang sold that little company off to 3M (Minnesota Mining and Manufacturing) for $2 a share. Mario had 20-timesed his money in eighteen months, and from $200 had $4,000. Jang suggested that he put that $4,000 into International Diagnostic Technology, another new firm that Jang was starting, and the new investment multiplied by more than 17 times, giving Rosati $70,000 from his original $200. He was made a partner after three years, before he was 30, largely because of all the start-up firms he had gotten involved with by then. His firm, now Wilson, Sonsini, Goodrich & Rosati, shares all new investments that it makes.

Rosati describes a typical start-up history of the present day. Roland Jang or someone similarly respected calls to say that an entrepreneur has appeared with what looks like a good idea. The two come to Mario, who offers to do the legal work for the new company. On the strength of Jang's reputation, Mario offers an investment by his firm, but never more than $10,000. Jang and the entrepreneur put in more, to show their commitment and bring the total for the new company to about $40,000. That investment is sufficient for the legal formation of the company and the preparation of its business plan.

Jang, with his wide acquaintance among Silicon Valley high-tech firms, helps the entrepreneur to find and to hire the key people he will need for the start-up. When the business plan is ready and the team is in place, Jang and Rosati introduce the entrepreneur to friends among the venture capitalists. Typically the new company needs a capital investment of $2 to $10 million. If the investment is made, the original investors do very well, at least on paper. In a recent case cited by Rosati, the original $40,000 was at a stock price of 10 cents per share. The venture-capital financing just thirty days later was at $4.50 per share—an increase of 45 to 1. The corresponding valuation of that new company was $9.5 million—an unusually high value for a start-up. Eighteen months later, the stock had split 4:1 and was going for $10 per share, so the founders' stock had increased in value by 400:1.

Usually at least a second round of venture-capital financing, and often two or three rounds, is required before a company can go public. In the later rounds, company valuation may reach $30 to $40 million, and if so, the "ground-floor" venture capitalists usually decline to invest, leaving the field to "mezzanine investors"—insurance companies and other institutions working through large funds. Those late-stage investors are hoping to buy in for one half or two thirds of the price that the company's stock will be sold for in a public market, perhaps six months later. Company valuation when the start-up goes public may be $150 million. The mezzanine investors will have made 50 to 100 percent on their investment within those few months—and they are the sort of conservative investors who traditionally earned no more than 5 percent.

In the 1950s and early 1960s, before venture capitalism as such was well established, the success of Silicon Valley depended on firms like Eimac, Varian, Fairchild, Intel and, largest of all, Hewlett-Packard. In the 1960s, a later generation of firms, such as ROLM and ESL, grew to large size. By 1971, when Rosati first observed the start-up scene, there were strong pressures within most successful companies to go public, so that investors and employees who had bought stock under option programs could sell, to obtain liquidity.

As Rosati describes it, there are just three alternative ways that the investors in a start-up company can make real, spendable money as opposed to paper. "Sometimes it's such a cash generator that you can just pocket the profits. But that's rare, usually just a 'Mom and Pop' firm—nobody else sticks around that long. The second way is to get acquired by a bigger firm that's publicly traded. And the third way, which is on the upswing now, is to go public."

As to the real chances of success for start-up high-tech firms, Rosati notes that the conventional figure is 2 in 10. But in his own experience the figure has been much higher: about 7 companies out of 10, he finds, ultimately make it. Five of those 7 are acquired, and 2 out of the original 10 make it big, like ROLM or Apple Computer. They go public. Mario cautions that the only way an ordinary investor, not in the know, can get in at the prepublic stage is through venture-capital funds. (There are now about 150 of those funds.) He recalls that when he first came to Palo Alto, in 1971, there were still only a few venture funds. Arthur Rock had been among the earliest of the venture capitalists. Jack Melchor and Burt McMurtry had founded the Palo Alto Investment Company. There was Mayfield I, with Tommy Davis and Wally Davis. By 1971, Burt McMurtry and Reid Dennis had formed Institutional Venture Associates.

Venture capitalism was then, as Rosati says, "a kind of club." One of the venture capitalists would see a good opportunity, he would call up his friends and they would finance the start-up of a company. Twenty companies per year was a lot. The entrepreneurs were seasoned men in their 40s and 50s. There was very little venture capital available, and it had to be carefully placed. A $300,000 start-up was big, and "You'd run around and tell your partners" about a start-up at $1 million. The investment money came largely from tax-exempt institutions. The Mayfield Fund had a good deal of Ford Foundation money. Venrock had Rockefeller

Foundation money. The venture-fund General Partners sold the investment, in Rosati's words, by saying:

> "You've been getting four to five percent in your traditional investments. We're getting explosive growth in the new 'sunrise' industries, mostly high-tech and mostly in Silicon Valley. You need a 'spark' in your portfolio. Give us just a little investment, at high risk, and we can be that spark."

Mario recalled that in the "old days" of 1971, the entrepreneur starting a company would sacrifice by taking a $20,000-per-year salary, and much of the early work would have been done in a garage. Now, he told me, entrepreneurs were being financed even before leaving their old jobs, while in their one-month termination period. And they were sacrificing little, if anything (except, of course, the security of a stable and successful established firm). The old fraternity of venture capitalists had weakened as the scale had increased. Now there were $3-million and $10-million start-ups. The large funds were splitting them up, taking 10 to 30 percent each to spread the risk.

He told me of the ranking of funds. There were the "A" string, the premier lead investors—Mayfield Four, Hambrecht & Quist, Brentwood Associates, Kleiner-Perkins and a very few others. Around them gathered satellite funds, the "B" string—well respected, but not on the network. And then the "C" funds—usually a family, or "guys who've made a few bucks on deals and gotten together." There might be foreign money in the "C" funds. They would be included in an attractive deal only if they had brought it to one of the "A" funds, or if $50,000 or $100,000 was needed to round out the deal.

Most of the foreign money being invested in venture funds was coming from France, Italy and the United Kingdom. Almost none was coming from Saudi Arabia, but a little was from Kuwait. None was from Japan. But Rosati observed substantial amounts of German and Japanese money coming for acquisitions, especially of medical companies:

> It takes five to seven years to develop a new drug, and that's too long for American investors. But the Germans think they're getting a steal when they pick up one of our little companies, like Roland Jang's IDT, for seven million dollars or so.

Mario Rosati had this to say about all that money being made in Silicon Valley, and how much of the personal wealth stayed there:

> The heavy hitters in the Valley? A hundred million dollars— that's unusual; count them on the fingers of one hand. Fifty million —maybe two dozen of those. Ten million—very acceptable. Five million—several hundred of that size. And almost anyone who's doing anything in venture capital has a million or two.

Rosati estimated that there were probably several thousand millionaires in the Valley. And San Francisco law firms were investing money there that had been inherited in the city. Most of the new wealth of the Bay Area was being generated in Silicon Valley.

That wealth is visible. Every hilltop near the junction of Page Mill Road and Interstate 280, the Junipero Serra Freeway "scenic route" along Silicon Valley's western edge, is topped by a new million-dollar house. After World War II, the Valley had still been filled with fruit trees in seemingly limitless orchards. Now the fruit is Apple Computer, a $500-million-per-year company employing many thousands of people. There are hundreds of other start-up firms not yet so well known, like Ibis Systems, Evotech, Compumotor. They are the reason for the new wealth of the Valley.

Whenever someone is acquiring wealth very quickly, onlookers tend to wonder about its legality. Obviously, the "heavy hitters" to whom Rosati referred could not afford to be involved in shady transactions even if they did not have moral objections to them. The investors and start-up companies must obey both state and federal laws.*

A corporation is a "person" in the eyes of the law. The stockholders of a corporation elect the directors, and the directors select the officers. All the stock that is issued by small start-up companies is "unregistered" stock, and there are four main restrictions on selling it:

1) The directors of the corporation must authorize the sale.
2) The stock must comply with federal law, the Securities Act of

---

* After my interview with Rosati, I learned as much as possible about those laws in the course of forming the Geostar Corporation—an action that took me from observer to participant in the startup scene. For others planning to form new companies, I suggest emulating the examples of Geostar and of many other successful startups by securing the services of a topnotch law firm, with securities-law specialists, at the beginning of the enterprise, in order to avoid legal problems before they occur. It is far more expensive to wait until a serious problem exists before calling in the lawyers.

1933. That says, in effect, "Either register your stock or find an exemption."

Registry costs $150,000, so the legal trend in recent years has been to approve more exemptions, in order to encourage capital formation. The traditional exemption is section 4(2) of the Act: "If the offering is not public, it need not be registered." The alternative is a "private placement." In a private placement, the company is allowed to sell stock not to the general public, only to "sophisticated investors." They are defined as people who are experienced in judging risky investments, and who have enough money so that they would not be affected seriously should the amount they risk in a private placement be lost. Their protection in any particular venture is knowledge, and a well-written stock prospectus for a new company, legally called a "Private Placement Memorandum," will list clearly every possible source of risk that the company's founders can think of. In a complex, nonstandard case a Private Placement Memorandum may be an inch thick.*

There is a further distinction, which often leads to hurt feelings when the friends of an entrepreneur offer to invest small sums in his venture. The Securities and Exchange Commission (SEC) does not limit the number of "accredited investors" who may buy stock in a private placement, but it defines "accredited" as meaning a net worth of over $1 million, or an income of more than $200,000 per year. And the SEC rules limit to thirty-five the number of nonaccredited investors who may be offered the chance to buy stock. Obviously, if the entrepreneur uses up those thirty-five "slots" on people who buy only a few thousand dollars' worth each, he has severely limited the total amount that can be raised for the new company.

3) The "Blue Sky Law" for the state in which the corporation is located demands in effect that the transaction be "fair, just and equitable" to the citizens of that state. (Its name comes from the notion of selling an unwary purchaser a "piece of blue sky.")

4) The offering must conform to the corresponding Blue Sky Law for the state in which the purchaser resides or in which he receives the offer.

In the fifty years since the Securities Act of 1933 was passed, the

---

* In the case of Geostar's second private placement, for $1 million, we chose also to go further than the letter of the law, and hold a two-day seminar for prospective investors so that all their questions could be brought out on the technical, legal, regulatory and marketing aspects of the new enterprise.

obvious loopholes in the law have been plugged by further provisions. It is illegal, for example, to buy stock in a private placement, then turn around and resell it to several people who resell it in turn to a larger number. That is called "pyramiding." When a company goes public, only the new stock that is then offered is legally registered. All stock bought earlier in private placements remains unregistered. It may be "leaked onto the market," but only with restrictions. The buyer must have held on to it for at least two years, and no buyer may sell on the public market at one time more than 1 percent of the total stock of the company.

For those who have bought stock, in private placements, in a start-up company that is later acquired, the manner of the acquisition is crucial. The worst alternative is acquisition by another company whose stock is not publicly traded, because then the original stockholders still have no liquidity. The next-worst is acquisition for cash, because the stockholders must continue to hold their stock for a year in order to receive capital-gains rather than direct-income tax treatment on it. Best of all is acquisition by a large company that is publicly traded, because then there is a tax-free exchange of the small company's stock for that of the larger, and there is immediate liquidity.

START-UP HISTORIES

The Machine Intelligence Corporation began as the brainchild of Dr. Charles Rosen. He had founded the Robotics and Automation Division of SRI International, and in the late 1970s was retiring from SRI to devote his time to ventures. The MIC "company" began literally in Charlie Rosen's garage.

It is a typical successful start-up, with a slower buildup and smaller multiples than those of the spectacular exceptions which Rosati described. MIC was incorporated in September 1978, with $100,000 raised by friends and associates. The company completed a first prototype of its robotic vision system by December 1979. Rosati introduced Rosen to Jang in 1980, and Jang brought in venture capitalists and helped to find a management team. MIC's second financing was about $700,000. A third, in 1981, was for $5 million, and was obtained from large venture funds. In late 1981, MIC had a champagne christening of its newest quarters, in Sunnyvale, for its more than 100 employees. At the time of that celebration, both Rosen and Jang were disengaging from MIC to take on still newer ventures. In 1982, a fourth financing was obtained. During those four years, the stock price rose slowly from $5 to $10, then to $25. A public offering would be made only after the firm became profitable. (As a leading venture capitalist said, "We normally finance the losses.")

GRID Systems, located in Mountain View hardly a mile away from MIC, is the concept of John Ellenby, an English economist turned manager. Ellenby directed the Xerox advanced-office-systems group until early 1980, when he left to form GRID. Ellenby's idea was to solve a universal corporate problem: Executives carry their work home with them at night and also travel a great deal. Whenever they are away from their firms, they are cut off from the corporate data base. He estimated that the total installed value of corporate data bases, hardware and software, in the United States alone was $400 billion.

The GRID concept would establish for each corporation a "Compass Central"—a hardware installation that would connect the corporate data base via the ordinary dial-up phone lines of the Bell System to any telephone handset to which an executive might have access, at home or while traveling, and from that to a "Compass" computer, a unit small enough and light enough to fit in half a briefcase. The executive would use it routinely at his office, and it would be no more obtrusive on his desk than a large book.

To give the Compass program updates and other services automatically, even when it was unattended, every Compass Central would be connected over a phone line to "Grid Central," a complex of nine IBM Series One computers at GRID's headquarters in Mountain View. From Grid Central there would flow to each Compass Central the latest in programs, program corrections and diagnostics. Whenever a Compass made contact with a Compass Central, its stored programs would be immediately updated and messages to and from the central corporate office would be transmitted. A balky Compass could even be examined and diagnosed via Compass Central from Grid Central, and if found to be out of order could be returned by Federal Express to the Mountain View office for exchange within twenty-four hours.

Ellenby's core group of about thirty experts in computer design and programming was drawn mainly from Hewlett-Packard,

Apple, Xerox and Data General. They designed the system and built the first Compass in just one year, finishing it in late 1981. By early 1982, they had raised $12 million in two rounds of venture-capital financing, and were checking out Grid Central. By July of 1982, GRID was expanding into a large, new set of three-story buildings at the edge of the Bay, and had about twenty of the Compass computers in test locations around the country. Interestingly, the investors in one start-up company often include the founders of earlier successful start-ups: Gene M. Amdahl of Trilogy Systems invested in GRID, as did executives from the Tandem Corporation.

Another textbook case history of a successful start-up is Seagate Technology, whose first public stock offering was made in September of 1981. Seagate was the creation of Alan F. Shugart, who formed its name from his own. He had become phenomenally successful as the first large-scale manufacturer of floppy-disk drives in the 5.5-inch size appropriate to personal computers. With Seagate, he intended to bring to the world of personal computers another technology once found only in larger machines: the Winchester disk. A Winchester (the name comes from a road at the northern edge of San Jose) is a rigid rather than a floppy disk. It runs steadily in a sealed enclosure, protected from dust and air pollutants. It achieves 25 times the data-storage density of a floppy by using "thin film" recording and pickup heads, made using the microscopic technology of printed circuits rather than by winding electromagnets with wire. Thin-film recording technology had been pioneered a decade earlier by IBM, but not widely used outside that company.

Seagate began development in October of 1979. Eight months later it had its first prototype of a 5.25-inch drive, storing on two Winchester disks 6.38 megabytes of data—as much as 50 floppy disks and with much faster access time. One month after that, Seagate began shipping to customers. Its first three months of sales totaled only $188,000, but its next quarter was $989,000, and by June 1981 its sales were $6.45 million. Seagate's legal work was done by Wilson, Sonsini, Goodrich & Rosati. On its board of directors there was Reid W. Dennis, whose own corporation was a General Partner of Institutional Venture Associates (IVA).

At Seagate's inception in October 1979, each of its founders had purchased $10,000 worth of Seagate's common stock, at half a cent per share. By June 1981, after the company had gotten well started through venture-capital investment, its stock was going for about $1.25 per share, and the firm began an incentive stock-option plan for employees, in which stock could be purchased only as long as an employee remained with the company. When Seagate went public three months later (just at the statutory two years for the sale of its unregistered stock), its offering was made at $10 per share, for a total of $30 million. Five hundred thousand shares, one sixth of the total offering, were sold by previous shareholders, for a total value of $5 million—a value 2,000 times as great as they had been bought for two years earlier. The offering was fully subscribed. L. F. Rothschild and Robertson, Colman each went in for about $5 million worth. Smaller amounts were bought by Paine Webber, E. F. Hutton, Kidder Peabody and New Court Securities, another arm of the Rothschilds. Shearson/American Express bought in, as did Merrill Lynch. Among the smaller buyers were the Banque de Paris et des Pays-Bas (Suisse) and the Vereins-und-Westbank Aktiengesellschaft. A company had come to life and grown, with international financial involvement, in just two years. And in the process, several hundred new jobs had been created.

On the involvement of foreign capital in American start-ups, and on the Silicon Valley phenomenon itself, Charlie Rosen of MIC said:

> There are always foreign investment people here on the Peninsula. They like the combination: high-tech talent around the universities; money available that's looking for start-ups; a ready market for high-tech products and finally the political climate of the United States, with no risk of revolution or radical political change.

In his view, Silicon Valley leads the world in that combination, and Route 128 around Boston is next. There are other places where the local business and political leaders are trying to put the same combination together: Minneapolis, Los Angeles, Austin, Atlanta, the Research Triangle in North Carolina and in Michigan around Ann Arbor, where the University of Michigan is located.

ORIENTAL INVESTMENT PHILOSOPHIES

Roland Jang was born in California, of parents already long established in this country, but in 1930, when Roland was a child, his parents sent him to China to get a "quality" education. When the Japanese invasion of Manchuria disrupted China, Roland was brought back to the United States, but he still carries a pleasing and ineradicable trace of Chinese accent. His Berkeley education gave him a strong background in electrical engineering and a Master's in chemical engineering. He then worked in various companies, and took part in the early development of Memorex. In 1972, Jang went out on his own, but after successfully starting IMT and IDT, he decided that staying through the two to three years of a full start-up was too slow for him. Since that time he has been midwife to the birth of Iris, Fermentec, Genus, MIC, Microbot and other firms. With a characteristic chuckle, he notes that "none of them has folded yet," so his batting average on start-ups is 1,000. Roland sees cultural differences between the Orient and the United States in regard to venture capital—indeed, there are differences between successful start-up entrepreneurs and successful managers in the United States.

There are four big differences, says Jang, between the U.S. and Oriental investment philosophies.

The first is ownership. If an Oriental invests a dollar, he expects a dollar's worth of ownership. But that may stand in the way of high-tech fast growth. In the United States, ownership is shared on the basis "How much money am I going to make back?"—on returns, not on investment. In America, according to Jang, the entrepreneur says:

> "I have a patent—sometimes not even a patent, just the idea for a product—and I'm going to work hard to build at least a ten-million-dollar company on it. So my idea and ability and hard work are worth ninety percent of the ownership of the company. You, investing a hundred thousand dollars now, deserve ten percent."

Easily done in the United States—*not* in the Orient.

The second difference is control. In the Orient, the investor wants ownership because he wants control. But in a high-tech start-up, he doesn't know enough to control right.

The third difference, says Jang, is tangibles. In the United States, we invest in high-risk because that's where the money is.

But Orientals want to see and touch what they're getting. That may be rational for real estate or hotels—but not for investing in potentials, in a business that may or may not be there in three years.

Fourth, banks, in the Orient, are much more available for investment, but favor old companies. So the closest thing they get to start-ups is within an old company when it initiates a new product line. "I think we'll see in the Orient the U.S. type of venture-capital start-ups," Jang says, "and we'll see it in Germany too—because they're smart people, no question. It will snowball after they've had three or four successes. It will come in as little as five to ten years." As to foreign money, he sees more Italian money than almost any other foreign kind. But he does not see foreigners trying to start new companies at home—"They bring their money here."

Why is it so much more attractive here for the individual entrepreneur?

> The great advantage the United States has over Japan or Europe is that here a man can take risks, lose a company, go through bankruptcy, start over and do well later. Sure it's better to win than lose, but there's no huge social stigma here in losing. But in Japan or Europe if that happens you're branded for life.

I asked about the difference between the "starters" and the "runners" identified so well fifty years ago by Nevil Shute Norway in his autobiography, *Slide Rule*.

"He's right—there is a tremendous difference, and you see it when the successful entrepreneurs leave a company after the start-up and go do another." Jang feels that the main reason a "starter" is not a good "runner" is that he gets bored by the modest 10- or 20-percent annual growth rate of a mature company, after the excitement and feeling of success that came with the early phases of doubling or tripling every year. If he stays too long he's likely to start failing, because he's used to taking big risks, figuring that he can intervene quickly to reverse a mistake. That may have been true when the company was small and only a few people's careers were at stake; but the response time for fixing mistakes gets too long when a company is big and public. Jang ended with a thoughtful comment that reveals his own motives as a nurturer of growth rather than as an exploiter:

I have two more things I want to do before I kick the bucket. One is to help a woman start a company. Two is to help a foreigner start one—the U.S. way, not the typical foreign way. How many companies start up in Japan or Europe that aren't U.S.-sponsored? Very few.

THE ECONOMICS OF VENTURING

Among the most authoritative venture capitalists, regarded unanimously as such throughout the Valley, are Burt McMurtry, Jack Melchor and Wally Davis. Few others are spoken of as highly.

Burt McMurtry's office at 3000 Sand Hill Road, in the hills between Stanford's Two-Mile Accelerator and Interstate 280, is in a typical Silicon Valley style: a two-story building with a large external staircase, roofed galleries, a red tile roof and much redwood. There are at least ten other venture-capital partnerships within a half mile of McMurtry's location.

McMurtry, born in Houston, took his degree in electrical engineering at Rice, then followed it with M.S. and Ph.D. degrees in the same specialty at Stanford. He developed microwave equipment for Sylvania, involved himself with lasers as soon as they were invented and soon managed a large R&D staff. In 1969—"the end of a great era in venture capitalism," as McMurtry says—Jack Melchor hired him away from Sylvania and the two formed the Palo Alto Investment Company (PAIC). A year later, McMurtry was made president.

PAIC invested $3.3 million in sixteen companies, of which twelve were start-ups. By 1980 those investments were valued at over $100 million, for an average 40-percent compounded annual growth rate for the investors. In 1969, $171 million in venture-capital investments were made in the United States. But in that same year, the U.S. capital-gains tax (CGT) was raised from 25 percent to 49.1 percent (Japan and Germany have no capital-gains tax at all). A drought struck venture capitalism in the United States as a result, and it lasted for eight years, as the table opposite shows.

"Liquidity is the major problem in venture capital," says McMurtry. "You need a public market to get liquidity, and when that market dries up, you see a throttling back in venture-capital money." The reduction in the CGT in 1978 happened, says McMurtry, because Dr. Edwin Zschau of the American Electronics Association (AEA) commissioned an independent audited study

Year	Venture Capital Investment (M$)	Comment
1969	171	CGT raised from 25% to 49.1%
1970	97	
1971	95	
1972	62	
1973	56	
1974	57	
1975	10	"oil-shock" recession
1976	50	
1977	39	
1978	570	CGT lowered to 28%
1979	319	
1980	900	
1981	1,200	CGT reduced further, to 20%

which showed that if the tax were lowered, the government would actually collect more total tax revenue.

The Treasury fought the change, argued that reducing the CGT would produce a huge shortfall in tax collections, but the congressional Office of Technology Assessment (OTA) backed it up. The study found that in the 1970s, a venture-capital investment of only $14,000 in a young high-tech company had created one new permanent, ongoing job, and the OTA calculated the total taxes on the investment and on that job. It was a very high payback ratio—far higher than that of any government assistance program in history.

McMurtry added that

venture capital is so extraordinarily productive. It's healthy. What we need from the government most of all is consistency. If we could count on the tax policy staying the same for ten years, I wouldn't mind living with the present twenty-percent tax; that's not onerous. It would be best if the government stayed out and let the system be driven by economics.

In 1973, McMurtry and Melchor, still close friends, parted to form separate ventures. McMurtry became one of the founders of Institutional Venture Associates, and says of himself, "No one else was crazy enough to try to set up a venture-capital fund in the depths of the '70s." But McMurtry and his three partners spent six months in 1974 raising $19 million, from six insurance companies and the Ford Foundation, which as a tax-exempt entity was unaffected by the CGT raise. That $19 million was a third of all the venture-capital dollars raised in the United States in that year. He then had ten years (the usual life of a venture fund) to put those dollars to work. Even through the 1970s, when the U.S. productivity growth rate was miserable, the companies in McMurtry's IVA averaged a 35-percent compounded annual growth rate, and were valued at $150 million by 1980.

"The major buildup in value for investors is the four-to-seven-year region in a fund," says McMurtry.

Since 1980, though—like the 1960s, but very unlike the 1970s—it's been possible occasionally to get huge increases in value in just eighteen months or two years. Apple Computer was one example: started in 1977, went public in 1980. Almost unheard of, but Seagate Technology, Inc., did the same thing between '79 and '81. Dave Marquardt [one of McMurtry's two partners in his newest fund, Technology Venture Investors], while working with my old friend Reid Dennis in Institutional Venture Partners, made a half-million investment in Seagate in May 1980. When Seagate went public in the summer of '81, that half million was suddenly worth twenty million.

Regarding the difference in decision making between operating a company and investing as a partner in a venture fund, McMurtry says, "Here you make relatively few decisions that really matter, but you're hung with those decisions, saddled for years." To be really well informed and available, McMurtry felt, a venture-fund partner could be responsible for no more than six to eight companies. An average start-up needed about $1.5 million, so in planning Technology Venture Investors in 1980 for three partners, he had calculated that his fund should not exceed $35 million. TVI is now investing at a rate of about $1 million per month, and with some reluctance McMurtry was raising his sights:

When we saw the flow of investment opportunity and the enthusiasm of our Limited Partners we decided to raise another thirty-five million, and we're adding two more people with the intention that they'll become partners in twelve to eighteen months. We're nervous going beyond three partners. Beyond six we'd be very uncomfortable. Funds get very big if you stay with them and keep investing ten or twelve million every year, but that may not be a good idea.

McMurtry chooses partners with strong backgrounds in technology: a B.S. or higher degree in science or engineering. He does so not with the expectation that he or they will understand everything the high-tech entrepreneurs are doing, but "so we're comfortable accepting that there's much that we don't understand. Comfortable asking stupid questions—we don't have to act smart." He prefers his partners to have M.B.A. degrees as well. In choosing, he looks for a peer, not a subordinate, because he wants partners who will keep him from "any tendency to hip-shoot in these euphoric times."

TVI works on a one-man, one-vote basis, although one of McMurtry's partners was brought in when barely past age 30. In setting up TVI, he wrote what he calls "the usual Superman specification" for his first partner. The partner had to be "younger than I, smarter than I, someone who'd been in venture capital in very tough times and had been in a lot of trouble—but had worked his way out of it." He wanted that partner's most recent job to have been in an operating capacity in a high-tech company, as a vice-president or higher, because "one simply forgets how difficult it is to manage people to make anything happen."

Jim Bochnowski met the specification and became McMurtry's first partner in TVI. Bochnowski had a B.S. in aerospace engineering from M.I.T., and an M.B.A. with distinction from Harvard. He had become a General Partner of Sprout Capital in 1972, and had achieved a compound rate of return of over 50 percent for the firms he had invested in, a list that included Shugart Associates. In 1977, Bochnowski left Sprout for Shugart, where he became the President and CEO. The company then had $13 million in revenues. Three years later, when Bochnowski left Shugart to join McMurtry, Shugart's revenues were up to $200 million.

But talented as McMurtry and his partners are, they don't take over management:

We won't invest in any firm that would let us do its management. Even if the entrepreneurs have never managed before, we want them to do it. Smart people will listen to input suggestions from everybody—including us.

One of the best reasons for going to a venture capitalist, Mc-Murtry believes, is that "he has analyzed you, he is on your side and he shares your objectives."

About 25 percent of the TVI partners' time is spent screening and reviewing applications for funding—potential deals. They spend another 15 percent of their time on communications among themselves—a fraction that McMurtry recognizes must increase as the number of partners goes up. The rest of their time is spent in educating themselves and in running the mechanics of the partnership. At the start of each new fund, time must be spent raising money. Says McMurtry:

> I have a very strong bias that every venture capitalist should have to raise money every few years, just to appreciate the problems of the people who come to him all the time trying to get investment money.

McMurtry's data on the ratio of venture-capital investments to opportunities come from his "deal log," in which his firm enters only proposals that the partners might think about seriously. About half the proposals made to TVI get discarded beforehand and never make it to the deal log. That process of instant discard is a plus for entrepreneurs. McMurtry feels that it is very important to give people a fast "No," and that stringing people along is unproductive.

From May 1980 to January 1982, TVI had logged in 575 opportunities. "Much above one per day," says McMurtry, "it becomes very strenuous." Of the 575, TVI had committed to invest in 19—about 3 percent. That was double what McMurtry had been used to from the 1970s. He also checked later and found a number that entrepreneurs should find reassuring: of the original 575 opportunities, about 100 were eventually financed by what he considered good sources. McMurtry dismisses the often-quoted figure 1 in 100 for the chance of getting financed. It is a statistic, he says, that venture capitalists generate only if they "leave a lot of garbage in their deal logs."

He finds that not only the quantity but the quality of investment opportunity has risen in the last four years. The reason, he believes, is that entrepreneurs are so smart and practical. If you're a bright technical guy and you know the chance of getting financed is only one in ten or twenty or fifty, you'll find better ways of spending your time. But at one in five, you'll try it.

The TVI partners spend a great deal of time digging into the potentially interesting deals and interviewing management, especially the top man. The digging is strongly personal: how good are the key people; how well focused are they; how well do they understand the market, the product, the technology? McMurtry points out that venture capital is a financial service in part, but that most of its activities are extraordinarily nonfinancial. As he says, "This is not a numbers business, it's a *people* business." Market studies don't help much, in his opinion. He very seldom does more than a brief cursory review of financial statements and the business plan until he has spent enough time with the management to decide that he is really interested. Together they dig into the numbers.

So much money is flowing into venture capital at this time that other funds are sometimes willing to invest on the spot, at first hearing. That puts pressure on McMurtry, but he finds that entrepreneurs are "very patient." At first they are flattered by capitalists who are willing to make instant "Yes" decisions, but then they think about it. Realizing that no one could possibly know enough about them to make an intelligent decision that fast, they often prefer to deal with McMurtry, and respect the care he takes.

I asked for examples of both success and failure. He gave me first the history of a failure:

> In 1970, with PAIC, we did a start-up with six people. I learned later that a large team is a red flag. The top guy, enormously talented, came out of a sales background with a strong technical component. Trouble is that a salesman spends a lot of time restructuring reality, not admitting to problems. After nine months of floundering, they asked me to run it.

McMurtry gave himself thirty full-time days, and found that there were only two people in the company with any management judgment. Both were women. One was a Phi Beta Kappa who was the Executive Secretary/bookkeeper. The other was a 21-year-old who was the production scheduler. Both were very bright. He made the Executive Secretary the acting General Manager and got her a good consultant. She got the company going and at least stabilized. But she came to him a year later—by then she was President—and said, "This isn't really going anywhere, but we've built value. Let's sell out." They did, in exchange for the stock of a publicly traded firm, and a year later that acquiring company went bankrupt. McMurtry's PAIC fund lost its whole investment.

As a contrast, McMurtry cited the ROLM Corporation (which his IVA had invested in) as an outstanding success.

ROLM is from the initials of its four founders: Richardson, Oshman, Lowenstern and Maxfield. The four, very bright young engineers who were in their 20s in 1969, started ROLM in that year. All were from Rice University, and McMurtry had recruited three of the four to Sylvania. Oshman worked for him for six years. They were, in his words, "classic entrepreneurs. They wanted to get into business, but without taking anything away from anyone. Wanted to create new wealth."

Jack Melchor financed the ROLM start-up. The four saw that the government spent $100,000 for a ruggedized minicomputer for military use, so they got a license from Data General, and engineered a rugged version of the Nova that they could sell to the military at a reasonable price. Then, in 1972, they went into the telephone PBX business. McMurtry calls it "a wild choice! That market had bodies strewn all over the landscape and lots of explicit competition." But the four thought they could build a fully digital Private Branch Exchange with plenty of value for customers. They invested the profits from the rugged minis in the new PBX product. The figures for ROLM after that decision:

Year	Revenue (millions of dollars)
1976	20
1977	35
1978	55
1979	121
1980	200
1981, 3rd quarter	100 (in one quarter alone)

McMurtry left the board of ROLM in 1981, not standing for reelection. That was an unusual move, as venture capitalists generally stay close to a winner. But McMurtry feels that "you have to re-pot yourself, get back with the funny little start-up companies." In the TVI portfolio there is Microsoft, Inc., the computer-software firm in Bellevue, Washington. A surprising entry was Pizza Time Theatre.

At first I shied away from it too, when Don Valentine offered me second- or third-round financing. But Jim Bochnowski said it was really high-tech, the world's biggest computer peripheral. He decided the growth rate looked good, we made a one-million investment and now it's worth six million after eighteen months.

That company provides for the under-10 set restaurants with animated games and troupes of Disneyesque computer-animated figures of animals that sing and dance.

Remembering McMurtry's policy of investing an average of $1.5 million (over time) in each firm he invested in, I asked about the low and high ends of the scale: the "little guy with an idea" and the projects, several of them discussed in Part II, that were highly valuable for society as a whole but too big and too long-term for venture capital. McMurtry observed that through most of the 1970s, the "bright guy with an idea" did *not* get financed. Most venture capitalists cannot afford the time, he feels, to start at that small a scale and put a team together from scratch.

The best source for the bright guy is individual investors, not funds. A former entrepreneur is best—someone who wants to get things going, and will put in his time and advice. Roland Jang is one of the very few who can and do play a pivotal role in real start-ups. I refer a lot of people to him. There's no one else that good except Jack Melchor, who is in a class by himself—a master, with incredible intuition about good young people with bright ideas. He works just closely enough with them, but not so closely as to keep them from building up their own experience. He's made a lot of money in a lot of things.

McMurtry had no clear solution to the problem of big, long-term projects that the country needs. He felt that the government did not do them well unless people were focused by fear, as in wartime. Without that, perhaps they could be handled by associations of venture-capital groups. I asked what troubled McMurtry within his own field of expertise.

A Japanese businessman said to me recently, "Venture capital in the United States is my best ally, because that prevents any U.S. company from gaining the large size that would give us Japanese real competition. It's too easy in the United States for someone to break away, taking a lot from a company." That bothers me. If there gets to be too much venture-capital money, investors may start pulling good guys out of growing firms so they can set them up in business. I hope I never do that. In recruiting, you pull one guy at a time and give room for promotion. That's very different from pulling the whole "XYZ Department" out of a firm and setting it up to do the same thing.

But also, I don't think Americans are at their best in big companies. The only gigantic U.S. company that's really effective is IBM —and it's run as a whole lot of closely coupled small companies. The others, like AT and T and the oil companies, are monstrously inefficient. You ought to be able to pull anybody really good out of one of those giants and set him up where he'll be a lot more productive. So I think the Japanese businessman is probably wrong. On the other hand, the American entrepreneurial spirit and ability is a great producer of wealth; the Japanese don't have it—yet, at least— and if I were they I'd be worried about it.

McMurtry's comment on "having to get back with the funny little start-up companies" was reinforced by the most senior venture capitalists, Wallace F. Davis and Jack Melchor. Although both have been very successful with venture funds of substantial size, both have chosen deliberately to concentrate on the small, embryonic phase of start-ups. They are cynical about the large amounts of money being invested through "super-big" funds by men far less experienced than themselves.

Wally Davis, born in 1918, took his B.A. in mechanical engineering and his M.A. in aeronautical engineering at Stanford. He spent most of the 1940s and '50s directing supersonic aerodynamic research in the Ames Laboratory (now NASA-Ames) at Moffett Field. In 1958, he founded Vidya, Inc., in Palo Alto. To raise money for it in that era when venture capital hardly existed, Davis visited the office in New York of the Rockefeller Brothers Fund (now metamorphosed into the Venrock venture-capital fund). Laurance Rockefeller, one of the earliest venture capitalists, had financed McDonnell Aircraft in 1943. With Rockefeller funding and the help of his partners, Davis built Vidya to a $3-million annual sales level over the next five years, and it was then bought out by the Itek Corporation. He remained as an Itek Vice-President until 1968.

In 1961, another Davis, Thomas ("Tommy," unrelated to Wally), had cofounded an early venture fund, Davis & Rock. That reached the end of its seven-year life in 1968, and soon afterward, John Wilson (of Wilson, Sonsini, Goodrich & Rosati) introduced the two Davises. By 1970, they had raised $3.8 million to launch the Mayfield Fund. In the early 1980s, the Davises managed not only the original Mayfield fund but three descendants, each of seven-year life and each larger than the last:

Mayfield I	1970	$ 3.8 million
Mayfield II	1975	8 million
Mayfield III	1978	30 million
Mayfield IV	1982	56 million

When a small start-up firm in Silicon Valley got overextended and was about to default on its payroll, its managers appealed in vain to banks and other funding sources. Finally they went to Wally Davis. He listened; said, "We can't let those employees not get paid" and advanced money to tide the firm over. With his help, the firm made a fresh start and became profitable.

On fund management, Davis remarked:

> The name of the game is to make good investments early, so that big multipliers can occur, the firms can mature and the investors will have liquidity by the end of seven years. But once you have all your money invested, you're out of business. So then you have to start up a new [venture] fund.

In March of 1982, when the Mayfield funds had a total of $80 million under management and Wally Davis had worked since 1970 to educate a new generation of managers, he resigned to form the Alpha Fund, aimed at seed investments and capitalized at only $12 million. Davis' copartners in Alpha were Brian J. Grossi, a young

Stanford-trained engineer, and Samuel Urcis, a founder of Geo-source, Inc., which had grown from its birth in 1972 to sales of more than $700 million per year in 1981. As usual with Davis, his motives were a mix of public spirit, practical self-interest and the sheer desire to have fun in his job.

In one venture, Alpha was putting in

just one hundred thousand to find a CEO and get the product planning started. Then we'll put in another two hundred thousand for the prototype and market test. The most we would ever put into one company is seven hundred fifty thousand to one million. That's labor-intensive for us. You don't have the time to invest a lot of money that way, so twelve million is more than adequate to do what we want.

As for the potential profits:

The prices are better in seed investment; there's more up-side to it. You go in at a better price, so there's a greater potential for big multiples. We like doing it. We're not bankers; we're all trained as engineers.

To save the General Partners' travel time for investigating prospects, Alpha concentrates on the Bay Area. It looks for investments in electronics and in biotechnology, fields in which Stanford University is strong. Alpha uses Urcis' knowledge of the oil business to find opportunities for technology transfer from Silicon Valley to Houston.

In the little town of Los Altos, characterized by Mission-style architecture with red tile roofs, stuccoed walls and courtyards of handmade Mexican tile, Jack Melchor maintains his headquarters. The largest staff group, about ten people, work on Melchor's "family affairs": real estate and investments that he has built up. A smaller number are assigned to Melchor Venture Management (MVA). Two people work on the Anglo American Fund, which the British Government asked Melchor to manage. It invests British tax money in the "Assisted Areas" of northern England, between the Midlands and the Scottish border.

Jack Melchor is in his mid-50s, tall, heavyset and with a manner that tends to be somewhat gruff. But often at the end of one of his typical statements (short, critical and uncompromising) his face will be transformed by a warm and charming smile. His B.A. and M.A. were in physics at the University of North Carolina, and he continued to a physics Ph.D. at Notre Dame. In one corner of Melchor's office there is a gold-plated shovel, which he has used in the groundbreaking ceremonies for a number of real estate developments in various Western states, as well as for such remarkably successful high-tech start-ups as the ROLM Corporation, Consolidated Video Systems and Biomation. Melchor's fortune is in the several tens of millions, and he has made it himself; his total inheritance was $13 from his grandfather.

Jack Melchor's first job was at Sylvania in California. After three years, he left to start his first company, Mel-Labs, in Palo Alto. It was a four-way partnership, which he found frustrating. By 1961, he had sold it to Smith-Corona-Marchant and entered into a joint venture, HP Associates, with Hewlett-Packard, to develop semiconductors and electroluminescence. It was so successful that it became an embarrassment for the larger firm; Melchor and his top executives were making more in capital gains than Hewlett-Packard's division managers.

In 1967, David Packard asked him to take over HP's old Dymec division and get HP into the computer business. Melchor did, and made the first computer priced at under $10,000. A medical problem in 1968, not serious but initially alarming, forced Melchor to "retire," and he decided to spend the rest of his life doing whatever he liked. That has been, for the most part, venture-capital investments.

I hired Burt McMurtry out of Sylvania in 1969, to form the Palo Alto Investment Company venture fund. We worked together for three years. I think I gave him ulcers, but he learned a lot, and he's done a superb job. I helped him make his first million. We're still good friends. In PAIC, we had three point five million invested for some offshore clients. By the end of 1980, it was worth a hundred and twenty-five million [a 38-percent-per-year compounded annual growth rate averaged over eleven years]. Since McMurtry's been on his own, I was a founding investor in two of his biggest winners, ROLM and Triad Systems. After thirteen years, my original fifteen-thousand-dollar investment in ROLM is worth about seventy-five million—the multiplier was five thousand to one.

Regarding his batting average in ventures, Melchor says:

One of the biggest mistakes I ever made was not getting in with Tommy Davis and Arthur Rock when they formed Davis and Rock and invited me to join. I didn't have the money then. That's the fund where they developed SDS and then sold it to Xerox for nine hundred sixty million.

About the ventures, I lose twenty percent of the start-ups outright, and write off a hundred percent of those as a tax loss. Another forty percent make from one cent to two dollars for each dollar invested. And the top forty percent are split. A third of them multiply from two to twenty times. Another third, you can count on twenty to a hundred times. And the last third, you can count on one hundred to five thousand times. It's a crapshoot. It's a reasonably good living. Everybody I've worked with on the Peninsula has made a lot of money.

Jack Melchor's choices for investment are made by his own Melchor Venture Management organization, which really means by Jack himself. MVA also advises the Portola Venture Fund, an offshore firm in Bermuda, with money from two British investors who wish to get out of the U.K. tax environment. The brochure for MVA and the Portola Valley Fund, intended for entrepreneurs looking for funding, shows a Leonardo da Vinci sketch, including his mirror writing, on its title page. It is a fund of only $12.6 million, about the same size as Wally Davis' Alpha Fund, and its venture philosophy is much the same as Alpha's.

Says Melchor, "I never go above a half-million investment in a start-up. It's a lot easier to get a ten-times multiplier at the low end, going from a hundred thousand to a million, than it is going from ten million to a hundred million." He added:

I don't go for the "management team" approach. I look for one bright guy who knows what in the hell he's doing. Roland Jang has the ability to go in and take over routine management. We don't do that. If we have to go in as managers, it means a guy we invested in just got fired, and I don't do that unless I have to.

Melchor followed Roland Jang into the Machine Intelligence Corporation, as the first outside investor after Jang. He frequently helps to identify and interview candidates for management in the firms he has invested in. He looks for board members who will really contribute, and believes they should have a financial stake. If necessary, he is willing to approve their being given that stake in the form of warrants (rights to buy stock at a specified low rate, like 5 cents instead of a going rate of $10).

The Osborne Computer Company was a start-up in which Melchor provided first advice, then capital. Melchor advised Adam Osborne to define one "right product" rather than going with his original idea of multiple products. Later, when Osborne had invested $70,000 of his own money, Melchor put in $30,000. I asked Melchor what it was worth now.

Oh, Christ, on paper about two and a half million, after two years. None of that means a damned thing in this business until you've cashed the check. There's a lot of inflated paper floating around.

Melchor's avoidance of large investments saved him from serious loss when the Osborne Company filed under Chapter Eleven a year after our conversation. He considers McMurtry more capable than himself to assess a company that is just prepublic, and notes that with the larger fund that McMurtry works with, there is no time to track the many small investments it would take to spread the money around in chunks of $150,000 to $300,000.

Some of the other managers of large funds also earn Melchor's respect, but he thinks there is far too much money now in the wrong kind of venture-capital funds, managed by people he has never heard of, who come out of an investment-banking background. Their style, he observes, is mob psychology. They don't make independent judgments, but instead try to tag along with seasoned venture capitalists with good track records—"the San Francisco Mafia," made up of firms like Hambrecht & Quist; Kleiner-Perkins; Caufield & Byers and a few others.

Melchor illustrates by a case history. A start-up company came to his investor group for a $2-million financing, where there was an appraised value of $4 million. Melchor offered the company part of it, and the managers then went to San Francisco for the remainder.

There, large banking houses insisted on throwing another $7.5 million into the deal, although "the guy running the company had never done a P-and-L statement in his life."

Melchor considers that kind of overenthusiasm fundamentally wrong. It occurs, he believes, because the bankers miss the point. They argue that they can get a better average return with such investments than with CDs. True—but in his view, the real purpose of venture capital is to build major companies with minimum investment.

One of the biggest deterrents to success, in his experience, is too much money too early. It "wrecks the first group of entrepreneurs, so they become big spenders, never learn how to run a tight ship, never learn to manage the bottom line." He regards the Genentech $35-million public financing as a case in point: it was "a great public-market deal—the timing was just right," but he doubts that at its assessed value of more than $100 million, anyone will come out of it in the next decade winning. When prices go through the ceiling, Melchor prefers to sit on the sidelines, so that when problems develop and more gullible investors come to him to bail them out, he won't have enough invested to justify a further effort.

Jack Melchor has done very well by keeping a cool head through the alternating fever and chill of the venture-capital cycle, and by knowing at first hand just what makes for success in a high-tech start-up. But for anyone with little knowledge and less financial backing who is tempted into taking a flyer on a start-up firm, it is worth noting that even Jack loses or breaks even on about 60 percent of the bets he makes.

Risky as venture capitalism is for the investor, it is extraordinarily productive for the country as a whole. It allows good new ideas to be turned into reality, rather than remaining as unfulfilled dreams. It stimulates and nurtures the creation of new companies in high-tech areas, where rapid growth is possible. That growth provides new jobs. And the people working at those jobs are learning skills that broaden their opportunities, preparing them for the decades of rapid change ahead. Every entrepreneur who leads a high-tech start-up takes a substantial risk with his career, but the accumulated wisdom and the financial support of venture capitalism give that gamble a chance of success.

Now, in the final chapter, we will move from Jack Melchor's heady, volatile, explosive world of venture capitalism to the other extreme in American industrial success. We will explore that $30-billion giant of corporate cool, IBM.

CHAPTER 9

WORLD-CLASS PLAYERS

To learn how to meet the challenge of productivity for economic competition, we have just investigated one extreme among American economic success stories: high-tech venture-capital start-up companies. Those examples teach us only how to initiate industrial growth, not how to sustain and build upon it. That is the domain of successful big corporations, and many of the people I interviewed, both in Japan and in the United States, felt that Americans were constitutionally unable to work effectively in such large organizations. If they are right, then there is no hope for America, because we can never match Japan's industrial might with a nation of cottage industries. But there are some very large American corporations which have structured themselves in such a way that their workers are as productive, effective and satisfied with their jobs as are the Japanese who work for Matsushita, Fujitsu and SONY.

We can learn the essential secrets of big-company success by studying three corporations.

The first, Delta Airlines, is in a service industry. As of 1982, it had operated at a profit for thirty-three consecutive years, and it was also the single most profitable airline in the United States. Its total earnings over 1971–1981 were $857 million—almost twice as much as those of the runner-up, Northwest Airlines. Delta also is known as an excellent employer. As of 1982 there was a backlog of 250,000 people wanting to be employed by Delta.

The second and third examples are industrial firms, both in the high-tech market arena of microengineering. The Digital Equipment Corporation (DEC) of Maynard, Massachusetts, is the smaller of the two. DEC was founded in 1957 and had annual revenues of $3.2 billion by 1982. From 1971 to 1981, its revenues grew by an average compounded rate of 36 percent per year, so that the DEC of 1981 was almost 22 times the size of the 1971 DEC. That rapid growth was not bought by risky borrowing or by acquisition. DEC carries almost no debt on its balance sheet, enjoys a double-A credit rating and has never acquired another company. Its growth was not at the expense of its employees, either.

It is hard enough to maintain rapid and successful growth for a company of DEC's size. But to do so when a firm is ten times larger still is phenomenal. Only one firm in the world has really succeeded at that level. It is the Colossus of Armonk, IBM—the third-largest industrial corporation, by my definition, in the world. Every year IBM grows by at least the equivalent of one DEC. IBM is also known as an excellent company to work for and enjoys strong employee loyalty.

The three companies share several characteristics: a strong ethical, moral philosophy; near or total absence of unionization and a corresponding sense of responsibility on the part of management to preserve job security by reassignment when necessary; promotion from within; and a consistent policy of open communication between management, staff and workers at every level. All three firms have built working systems of mutual trust and responsibility that result in high morale, productivity and quality. And the two industrial firms among the three have deliberately structured themselves as collections of semi-independent smaller "companies," whose individual managers have a great deal of authority. Many of those characteristics are what it is now fashionable to call "Japanese," but in all three companies they were developed long before American firms began looking to Japan for guidance. The bottom line is: they work.

Delta Airlines is an interesting example precisely because it must use exactly the same technology that is available to all its competitors. Its one big advantage is its corporate philosophy and its freedom from unionization. Except for its pilots and a few flight dispatchers, Delta's work force is nonunion. It offers qualified employees opportunities to change job assignments for personal advancement. That raises productivity, and no one at Delta ever says

"That's not in my job description." Rob Buck, a young copilot at Delta, said, "If I see a piece of luggage that's fallen off a conveyor, I pick it up and put it back on again. If I see a ground crewman struggling with a heavy piece of equipment, I give him a hand. On most other airlines, that would lead to a union grievance or possibly a strike that would shut down the whole operation."

Delta has a strong sense of responsibility to maintain full employment. Even when the oil-shock crisis of 1974 made the airline cut back its flight schedule, surplus pilots and flight attendants were kept on the payroll, assigned to different jobs. For the past twenty-five years, Delta has never furloughed a full-time employee for economic reasons. Except for some employees, such as pilots, who must be highly trained before being hired, Delta makes a practice of bringing in its employees in entry-level positions and promoting them according to their accomplishment on the job. Turnover among the employees is remarkably low. Despite many attempts over the years by union organizers, Delta employees have not even bothered to vote during the past several years on whether to unionize.

DEC's President, Kenneth Olsen, has a career history that would be normal for the top man in a Japanese high-tech firm, but is unusual in large American companies: he was educated (at M.I.T.) as a computer engineer and has stayed with DEC since he founded the company in 1957. Though DEC is far larger than Fanuc, Ltd., Olsen's career is remarkably similar to that of Seiuemon Inaba. Like Inaba, Olsen pursued a unique technical goal—the development of a minicomputer directed from a terminal. The first mass-produced realization was the PDP-8, marketed in 1965 and still going strong as the heart of the DECmate word processor. In 1969 he brought out the PDP-11, a larger minicomputer with a remarkably effective architecture. More than a quarter of a million PDP-11s have been sold. In the late 1970s he introduced the VAX-11-780, a revolutionary new 32-bit supermini with the added feature of "virtual memory." The VAX architecture allows the user to work with it almost as if its fast memory were limitless, because the machine automatically calls upon disk drives and tape drives to store program and data materials that are not called for often. The VAX series was later extended downward to the smaller VAX-11-750 and, in 1982, to a VAX-11-730 that cost hardly more than a word processor.

Like Japanese firms, DEC is oriented toward growth and market share rather than toward profit skimming. It has never paid a dividend, but instead has plowed back profits into expanding at its 38-percent-per-year rate. Olsen finances DEC's growth conservatively, by selling stock rather than by borrowing. His Financial Vice-President, Alfred Bertocchi, remarks, "We want to be able to get debt financing whenever we need it—and then never need it."

Olsen has organized DEC into eighteen business units, each largely autonomous. To keep his engineering staff of five thousand active, he subdivides it into small teams, with about thirty engineers in each. Realizing that the creativity of his engineers is DEC's most priceless asset, he does not push them into management jobs, but instead often pays them more than their managers. They are free to work whatever hours they choose, preserving an M.I.T.-ish academic flavor in the research laboratories. The scholarly approach is maintained by DEC's extensive educational department, which is constantly occupied with technical teaching and management instruction. To supplement that, DEC usually has twenty or thirty of its executives off on leave taking courses at university business schools.

In its worker/management relations, IBM is unique in its size range. From the company's inception, it has softened the traditional adversarial worker/management relationship. While some of its foreign corporate operations are unionized as necessary to satisfy local laws, the U.S. IBM is the largest wholly nonunion company in the world, with more than 250,000 employees. Later, in conversations with IBM's CEO, we will find how that has come about and what reciprocal responsibilities it entails.

LABOR UNIONS
Delta Airlines and IBM suggest that we need to reexamine the role of unions in American industry, and whether they have a place in the 1980s and beyond. They were formed, decades ago, when working conditions were often poor and dangerous, and exploitation of workers by management was commonplace. But industry-wide unions grew to enormous power in a protected environment when U.S. industry was virtually free of foreign competition. That environment no longer exists. When unions were building their power base, the main issue was: are workers being denied an ade-

quate share of the wealth produced by their labor, and are their working conditions dangerous and dehumanizing? Now the safety of the workplace is enforced by strict federal laws, and the chief human issue has become: is the worker going to lose his job because the firm he works for cannot sell its products in competition with Japanese products of higher quality, produced at lower cost? Striking, the traditional union weapon, actually aggravates rather than helps to solve that problem.

Over the past decade, total membership in U.S. unions has declined significantly, which suggests that many workers no longer feel that unions are helping them. But the adjustments that will have to be made by American workers to remain employed are far more drastic than many are ready to accept. In the steel industry, for example, U.S. labor costs are 91 percent higher than those in Japan, and in the auto industry about 67 percent higher. Tight, inflexible job classifications that were tolerable when technologies changed slowly have become destructive now that the rates of change are explosive. Those tight classifications, some of them frozen even into the laws of states and cities, are particularly detrimental in the construction, railroad and airline industries.

It was ironic that in 1982, a time of virtual stagnation in the construction industry, with massive unemployment, union carpenters in California demanded and got 10-percent hourly wage raises. In the auto industry, even after 1 million U.S. jobs were lost between 1978 and 1982 as a result of foreign competition, UAW workers at GM voted by the narrowest of margins, 52 percent to 48 percent, to allow concessions in work rules and hourly wages. In the 1980s many U.S. executives and many U.S. workers are willing to discuss gain sharing, in which company cost records are opened as they have never been before, and in which incentive wage payments are dependent on productivity increases.

The corporate philosophy of IBM was set by one man, Thomas J. Watson, Sr., just as the philosophy of the Matsushita Electric Company was set by Konosuke Matsushita. Those philosophies are remarkably similar. The American firm began operations in 1914, just prior to World War I, and the Japanese just after its conclusion in 1918. IBM was formed by the consolidation of three firms that were "high-tech" by the standards of the late 19th century. Dr. Herman Hollerith, of the U.S. census office, had invented in the

1880s a series of machines that performed counting and adding operations on data obtained by sensing electrically the pattern of holes in punched cards. His machines cut the tabulating time for the 1890 U.S. census from seven years to three. In 1896, Hollerith formed the Tabulating Machine Company, based in Washington, D.C. Eight years earlier, a jeweler, Willard Bundy, had invented a mechanical time recorder. By the 1890s, the International Time Recording Company was established in Endicott, New York, to market his invention. In 1891, the Computing Scale Company was formed at Dayton, Ohio, to sell a "smart" scale devised by Julius E. Pitrat. The three firms merged in 1911 as a New York State corporation, and came under single management in 1914 as the Computing-Tabulating-Recording Company (C-T-R). Tom Watson, Sr., joined C-T-R in that year as General Manager, and became its President a year later. The corporate name was changed to International Business Machines (IBM) in 1924.

At the close of World War II, the senior Watson assumed that the total market for electronic computers would be less than 50 machines. His son Tom, Jr., however, sensed the coming explosion of information technology. His ascension to the IBM helm in 1952 set IBM on the new track, and the company never looked back.

A decade later came an equally important decision point. IBM was becoming so big that its unrelated products, running incompatible software, were getting in each other's way in the marketplace. The solution, heralded as the "You bet your company" decision, was to replace the entire line of computer products by a single new series, the IBM 360.*

By 1980, IBM was doing a $26-billion annual business.

A year later, the U.S. data-processing industry as a whole employed a million people and produced $60 billion in revenue (nearly half of it by IBM), while earning a record $7-billion trade surplus. IBM's budget for R&D was twice as large as that of the National Science Foundation. The company spent as much on microelectronics R&D in 1981 as did the entire nation of Japan. It applied high technology in a homely application with its 3687 scanner, which used holographic "wraparound" laser light to scan the bar codes on packages held by supermarket checkout clerks.

---

* The number meant to imply the 360 degrees of angular coverage in all directions. To go the commercial division one better, IBM's Federal Systems Division called its military computer, later used in the Space Shuttle, "4-pi," to imply coverage over a complete three-dimensional sphere.

At the exotic high end of the product scale, the company introduced the 3081 computer, which employed two central processors working in parallel at a cycle time of 26 billionths of a second. The new machine was built up of "Clark boards"—hand-sized modules made up of heavy steel shapes in which spring-loaded pistons pressed on each of 118 computer logic chips, forcing them against a metal block cooled by chilled water. Fast-moving helium gas assisted the heat transfer. Sixteen Clark boards containing 750,000 circuits were packaged within 4 cubic feet in the 3081. (Such a high packaging density is essential in pushing as close as possible to the speed limit set by the velocity of light.)

I questioned John Opel, IBM's President and CEO,* about an allegation by Fujitsu that its own competition to the 3081, the FACOM M-382, was superior in that it got away with air cooling. His reaction was vigorous:

> They're dumping the heat problem in the customer's lap that way, using *his* air-conditioning system. Water cooling is far more efficient thermodynamically. We have the highest packaging density in the world with the Clark boards, for highest speed, and nobody else is even close to that technology.

IBM:/COMPUTER CHIPS

In 1981, IBM went into experimental production of the world's highest-density memory chip, storing 288,000 bits, and it did so using its regular assembly lines. The company was wary about revealing its automated assembly techniques, but did issue publicity photographs of an electron-beam machine, which used a cutting beam of electrons only one ten-thousandth of an inch across to tailor the interconnections on 2,000 high-density chips every hour. In its Yasu plant in Japan, an automated materials-handling system much like that of Fanuc, Ltd., was used to distribute 55,000 different parts to the production lines.

IBM's management understood, and emphasized in all its internal communications, that the company would have to maintain technological leadership if it was to succeed in its ambitious goal of growth equal to or greater than the industry average of 20 percent per year—a goal that would make IBM a $100-billion-per-year company by 1990. The "firsts" of its earlier days, such as the vac-

* Opel was made Chairman of the Board of IBM in 1983.

uum-stabilized tape drive, magnetic disk memories and the mass production of 64k memory chips, would have to be succeeded by still newer innovations that would be copied, in their turn, by smaller companies.

As of the early 1980s, IBM seemed to be maintaining its lead. In addition to the Clark-board technology, with twenty layers of printed circuits in each module, the company developed a new thin-film recording head, made by plating and photolithography—the same technology, pioneered by IBM, that had been adopted by Silicon Valley start-ups like Seagate. The IBM version was "big iron," a disk file spinning nine disks at 3,600 rpm on a spindle, to store 1,261 megabytes of data with an access time of only sixteen milliseconds. To achieve that density of information (two mega-bytes on an area the size of a fingernail), the IBM recording and pickup heads floated just 10 microinches above the speeding disk surfaces, tracking each microscopic hill and valley by sensing and feedback, like a fighter plane following the contour of terrain. That spacing was only 5 times the average distance traveled by one air molecule before colliding with another.

Opel and his colleagues realized that achieving their goal would require IBM to compete in every possible market arena, rather than selecting only the easier ones. By 1982, there was speculation that IBM would enter the "chip wars" directly, selling a part of its chip production to other original-equipment manufacturers (the "OEM" market) rather than holding production to just the levels needed for its own production lines. As Paul Low, an IBM Vice-President, said, "We contemplate it regularly." The results would be dra-matic.

Both IBM and Western Electric were large-volume producers of chips. As of 1982, IBM alone was producing more 64k memory chips than all the world's other semiconductor firms combined. Its plant at Fishkill, New York, was the largest single semiconductor factory in the world. And the IBM production lines were achieving a phenomenal 60-percent yield of good 64k RAM chips, compared with a maximum of 50 percent claimed by Japanese producers. Such chips sold to OEMs at that time for about $5, and the market for them was expected to reach $2 billion by 1985, out of a total projected world semiconductor market of $27 billion.

IBM's entry into the chip wars, if it occurred, could be with products still more advanced than the standard 64k memory chip.

It might be made with the 288,000-bit chip, or it might be made with a quarter-inch chip, announced in 1982, which stored 72,000 bits of information in a way that permitted their retrieval in only 50 billionths of a second. But if previous policy was followed, OEMs would not get their hands on IBM's most advanced components. The company began selling its Model 680 disk assembly to OEMs in 1982, for marketing under their own brand names, but the 680 did not incorporate IBM's thin-film technology.

IBM:/COMPUTERS

There is a unique title within the IBM headquarters staff structure, that of "Chief Scientist." Since 1972 it has been held by Dr. Lewis M. Branscomb, a physicist (Duke University, summa cum laude, 1945, followed by M.S. and Ph.D. degrees from Harvard), who left the directorship of the National Bureau of Standards to join IBM. His job description, like that of Dr. Kitsuregawa at Mitsubishi Electric, is vague: he is responsible for "guiding the corporation's scientific and technical programs to ensure that they meet long-term needs." When he took the position, his predecessor, the universally respected physicist Emmanuel Piore, had only one piece of advice for him: "Don't ever let them put you on the organization chart."

Computer capability—or, as Branscomb says, "raw compute-power"—is measured by "benchmarks," the time and cost associated with processing a given mix of computational tasks. For one such mix, consisting of 1,700 typical data-processing operations requiring a total of several million computer instructions, Branscomb's staff had summarized the times and costs over a period of several decades for the IBM computers of each era:

Year	Time Required	Cost
1955	375 seconds	$14.54 per benchmark
1960	47 seconds	2.48
1965	29 seconds	.47
1975	4 seconds	.20
1982	1 second	.07

In just twenty-seven years, computers had progressed so far that they could do the benchmark tasks nearly 400 times as fast as in

1955, and at one two-hundredth of the cost. Adjusted for inflation, the cost figure approached one thousandth. The rates of change were still being maintained in the 1980s, with computational power increasing at about 40 percent per year and the cost for the bench-mark tasks decreasing at 25 percent per year. Over the same period, the reliability of computers had improved 10,000-fold.

Branscomb pointed out that by sometime in the 1980s, given the continually rising costs for the services of people, it would become economical to have programs written by other programs, even if those superprograms were complicated and lengthy. The cost of computing at 1 million instructions per second for an entire month would fall to less than the monthly labor cost of one human pro-grammer. The trend was evident, he felt, because through the help of sophisticated programs, the productivity of the software writers was already increasing at about 10 percent per year. It would have to increase a lot more.

The investment of a typical company for data processing in the 1950s had been 70 percent for hardware and only 30 percent for software (programs), but in the late 1980s the emphasis would be reversed, with the hardware absorbing only 15 percent of the data-processing investment and the software taking 85 percent. One of Branscomb's concerns was that corporate management might not be quick enough to realize the domination of software over hard-ward in future systems.

The linkup of computers to form networks interested Branscomb a great deal. IBM used two internal networks. One, implemented top-down with tight controls, took care of processing orders, pay-checks and other sensitive data. The other, VNET, was by his estimate the largest private network in the world. It had 600 proces-sor nodes (computational focal points) in 110 cities among 18 coun-tries. It had grown bottom-up, built by the laboratory engineers and scientists who needed to exchange information with each other. In VNET, a still-growing organism, flexibility and ease of use were prized above efficiency and absolute reliability.

While high on computer networking, Branscomb was scathing on the limits imposed by the conventional telephone system. Two computers trying to exchange data over narrow-band telephone lines were, he felt, "like two lovers trying to communicate through the post office." By contrast, communications satellites had suffi-cient band width for computers to communicate through them at the same speed at which their CPUs operated—and the cost per channel of satellite communications was falling at 40 percent per year. But in the long run he saw fiber optics as the method of choice for high-speed transmission of data.

Branscomb's current interests are focused on three interrelated areas: technological development; the economic survival of the United States in competition with the rest of the world; and American education. In the IBM laboratories at Yorktown Heights, "fragments of circuitry" based on the quantum-mechanical Josephson-junction principle have been made to work at a switching time faster than 20 trillionths of a second. As Branscomb says, that is going to lead to a computer that is "very small, very fast, very cold and—at least initially—very expensive."

Cramming all the functions of a complex computer into such a tiny space will require advancing printed-circuit technology, as well as using quantum-mechanical effects at temperatures near absolute zero. Today's "conventional" silicon-chip technology is approaching the limits of photolithography, the optical technique by which circuits are "printed." Printing in that sense means optically exposing a prepared blank board to a pattern of light and shadow. Where the light falls, a photosensitive chemical coating on the board is altered, allowing it to be etched away by an acid. After etching, only the current-carrying conductor lines, corresponding to the black areas on the original photograph, remain. In the early 1980s, conventional step-and-repeat optical devices could produce circuit lines just 2.5 to 3 microns wide—close to the limit of that technique, which is about 1.0 to 1.5 microns.

To go smaller will require, in Branscomb's opinion, the use of hard ultraviolet light, X rays or electron beams. The limit posed by the structure of the silicon crystal itself will be reached at about 0.25 micron. In that area of development, the engineers pushing technology for peaceful uses rub elbows with those working on the DOD VHSIC program (described in the "Microengineering" chapter). IBM's work on the VHSIC program is done at its Manassas, Virginia, laboratory. Branscomb would prefer that the DOD pursue its own goals without muddying the waters by arguing civil benefits for the VHSIC program.

Finer scale in the circuit lines has made it possible for specialized semiconductor firms like Intel to produce VLSI "micromainframe" computers on a chip. Branscomb confirmed that IBM is devoting efforts to micromainframe development. He said that because there were twelve different major architectures in the IBM computer product lines, there were many programs, in the IBM research division and in other divisions, leading toward micromainframes. The 370 is an old architecture, not easy to miniaturize, but it will be put on a one- or two-chip micromainframe. And a small machine, the VM/CMS, will be packaged that way also.

IBM believes a micromainframe can be especially useful if embedded in a new computer as a coprocessor, to translate software previously written for an older machine. Branscomb said that IBM was particularly interested in a one-chip micromainframe Series/1 computer, because the Series/1 embodied a modern and uniquely useful architecture.

That machine came out of an IBM research laboratory, but was later adopted by management as a standard item in the product line. It is noted for its ability to carry out many tasks simultaneously, and is frequently used as the communications link between larger computers. A Series/1 is at the heart of IBM's Audio Distribution System. Others manage energy in large stores and factories, for highest efficiency. Still others control videotex data systems. Large customers like airlines can install a "ring" of up to sixteen Series/1s, separated by distances of as much as a mile, with each computer in the ring located at a data-processing work area: a passenger-information source, a baggage office, a freight office and a flight-plan center. Each Series/1 in the ring can direct information to any or all of the others at speeds up to 2 million bits per second.

The point of all the precision in microengineering is to process more information and to do so more quickly. Information is a peculiarly valuable commodity, because it is not consumed by use. In fact, consumption usually increases its value. No natural law limits how much of it people can have. One problem, however, is that many individuals are information-poor, in the midst of an abundance of information.

The amount of information storable for a fixed dollar cost as binary computer data (which means, in principle, just about every kind of information there can be) is increasing at a compounded rate of 35 percent per year, so the total data base per dollar can be 100 times as large in the year 2000 as in 1985. Even now, all the information in the Library of Congress (about 20 million books) could be stored in just twenty of the IBM 3850 disk files. Branscomb foresaw a distant time when optical-fiber technology might approach the theoretical carrying capacity of the visible-light laser, and at that point the total information in the Library of Congress could be copied and transmitted in just one fiftieth of a second. Even at today's level of technology, all that information could be transmitted over a single optical fiber in only three weeks, and Branscomb was sure that the electronic storage and transmission of information would replace most of the specific-message mailing system relatively soon.

ERGONOMICS

Branscomb thinks the key software developments of the remaining years in this century will be in making computers easier to use—more "user-friendly," as it is advertised frequently. There is a growing art, "human-factors research" or "ergonomics," that focuses on smoothing the communications between people and their machines. He believes that in the 1980s the improvements will be mainly for engineers, scientists and other specialists; in the following decade, the general public will benefit. Others feel that the intense pressure of the public small-computer market will bring the new, easier systems to the personal-computer arena at least as soon as they reach the more specialized, lower-volume markets. In either case, the developments will help the general public to find that the computer is, in Branscomb's words, "a powerful servant—not, for all our hyperbole, a powerful god."

At present, there are still distinctions between information in the forms of numerical data, audible speech or music, and visual images. It is Branscomb's opinion that new advances in hardware will soon erase those distinctions, so that a single medium of storage or of transmission can include all those forms of information. The precursors are already here. Early in the 1980s, IBM introduced its "Audio Distribution System," in which binary data could be routed to any destination by an executive's key strokes on an ordinary push-button telephone. Britain, Canada and West Germany are particularly active in a more sophisticated form of information transfer, "videotex," in which a video terminal assists the executive in his search for data and his selection of them for transmission. The PRESTEL videotex system in Britain was already usable by 1982, and Branscomb thinks that the more powerful TELIDON system developed in Canada will become a new standard.

Friendliness to the user means dealing in his own language and modes of expression rather than in an abstruse computer jargon. Branscomb thinks that by the mid-1980s, computers will be able to read our handwriting—not just the specific printed typescripts that they can read today. And they will be less forbidding than they are today. The ergonomics pioneer Alphonse Chapanis of Johns Hopkins University found that people cooperating to solve a new, unfamiliar task did so most effectively when they communicated by fragments of speech, not by complete sentences with correct syntax. Branscomb concludes that the computer of the near future will test our mood with a little sociable chitchat before it gets down to serious work, and even then will accept conversational fragments with key words and phrases, rather than insisting that we get the punctuation exactly right. It will send less authoritarian and arrogant messages than "SYNTAX ERROR," or "ILLEGAL PASSWORD" or "FATAL ERROR, RUN ABORTED" when we make mistakes; it will be more tactful. (Unfortunately, the primitive attempts made so far at programming of that kind are so insufferably arch as to be exasperating rather than encouraging.)

RECOGNITION EVENTS

While these technological opportunities give IBM lucrative new markets to penetrate, they do not give the giant company any special advantage. They are based on physics that is universal and accessible to IBM's competition. It helps to have a head start; IBM was well up the S curve of growth before many of its present competitors even existed. But as the U.S. automobile industry demonstrated, reaching world leadership can be a prelude to a fall unless, despite success, a firm continually innovates and maintains enthusiasm. IBM has avoided traps that usually go with bigness and success because of its unique attitude toward people.

In IBM's appreciation of its top-performing employees, "Recognition Events" play a central role. Those events go back to the 1920s, when the elder Tom Watson brought his leading salesmen and engineers together in annual tent camp-outs at Endicott, New York. The central theater for each event is still called "the Main Tent," though it is now generally the grand ballroom of a leading hotel.

Great care is taken in the planning of Recognition meetings, and for the value received, it makes sense. The Recognition Events are important in motivating employees to win and deal effectively with customers; the return on investment is very high.

In 1982 the Recognition Events were of special importance, because in that year IBM carried out a massive reorganization. Until 1982, IBM's divisions had corresponded to product lines: office products, data processing and so on. But the explosive rates of technological change were blurring the old distinctions; a single computer could now do scientific calculations, keep track of employee payroll accounts and simultaneously be used as a word processor for the typing of letters. As a result, IBM salesmen from different divisions found themselves knocking on the same doors. In the reorganization, the customer rather than the product was taken as the key, and once a salesman had made his contact, he could sell any machine or any program made by IBM. During the Recognition Events, John Opel and other top executives spoke to assembled IBM-ers in several cities to give information and answer questions about the reorganization and its impact on IBM's future.

THE CEO

John R. Opel, President of IBM since 1974, was made Chief Executive Officer in 1980. In 1983 he retained that office, while leaving the presidency to become Chairman of the Board. Opel's career includes education in science and management and a long apprenticeship in different aspects of the company's business. He was born in Kansas City and grew up in Capital City, Missouri. He went to Westminster College and the Illinois Institute of Technology, majoring in engineering and ceramics chemistry. He followed that with an M.B.A. in 1949 from the University of Chicago, and then immediately joined IBM. Opel worked in sales with a farmers' cooperative in Missouri, and in the early 1950s moved to Washington to be IBM's technical representative to the U.S. Air Force on the logistics of the Strategic Air Command. (The old catchphrase "IBM means 'I've Been Moved' " took on a slightly different meaning when I learned from other IBM-ers that moves are usually promotions, so that their number is a measure of status.)

Opel's next job was in product and financial planning, and in 1957 he was put in charge of all the technical writers in IBM. In 1959, after a tour of duty in the training of salesmen and sales engineers, he was tapped to be the assistant to Tom Watson in IBM's corporate headquarters. The company tested him for public-relations ability by making him Director of Communications and then made him Group Director of the Product Line.

It was a time of controversy and rapid change. There were many different architectures, and new types of memory were in competition: disk, core, tape. Designs were constrained by the paramount need to conserve fast memory, which at that time required the hand threading of ferrite cores with fine wire, one core for each bit. Ultimately, the sweeping decision was made to adopt the 360 architecture. It was risky, but it worked, and Opel was at the heart of it.

When asked to identify IBM's uniqueness, Opel's answer was in terms of people, not hardware:

> The "Open Door Policy" works. If an employee walks in my door and says he's been mistreated, I take it at face value. I start with the premise that he's right, and the burden of proof is on his supervisors. The most misunderstood aspect of IBM is how good the company is at finding and keeping the best people. Tom Watson used to say, "Take away the factories, take away the machines, but leave me the people, and I'll start the company back up again very soon." The nature of our work helps us a great deal, also. We're embedded in a business that's forced to change and grow, so we have full employment and a constantly increasing level of skills.

Opel pursued the uniqueness of information:

> You never get satiated on information; there's always more demand for it, and it has cash value. Remember, the Rothschilds got rich on it: they arranged for carrier pigeons to be released from the battlefield of Waterloo as soon as the outcome was certain, and they made their fortune by the news.

Opel's comment supports the interpretation that technically advanced nations are going through a demographic shift that will transfer more and more people into information-related jobs, out of direct factory production. Steve Haeckel, of IBM's corporate headquarters, said that by 1960, 40 percent of all American workers had jobs related to information, while 35 percent were employed in our industries, and 5 percent were still engaged in farm work. By 1982, 53 percent of American workers had information-related jobs, and the factory work force had shrunk to only 20 percent. In agriculture, automation was almost complete, and only 2 percent of Americans were needed to do agricultural work. If industry goes the way of agriculture, the nation will have nearly two thirds of its workers dealing with information, while less than a third will be in service occupations and only about 7 percent will be in manufacturing and agriculture.

IBM's need to compete across the board in order to retain market share as the information industry grows still further led Opel to analyze decisions on entering new markets, and on making or buying components:

> The key is to get the numbers right. You've got to know the costs and the value-added for each channel of sale. There's no point in investing finite resources in low-yield products. As for make/buy, in a vertically integrated firm like ours, there are profound decisions to be made all the way back to the raw materials. A good example is copiers: there's not much to invent, because it's a standard selenium photosurface. So there's a continuing price war, and IBM didn't do well in that market. But for the organic photoconductor, where we have an IBM proprietary technique, we have a good opportunity. We'll put R-and-D funding on the unique things that keep us ahead. When a technology gets to be mature enough that lots of manufacturers can build it, we'll buy from the low bidder.

Concerning robotics, Opel said, "We expect robotics to be substantial—maybe seven hundred million dollars per year. But not big for a forty- or fifty-billion-dollar company."

COMPETITION

The size and the success of IBM have made it a target for attack both by its competitors and by the Federal Government. Between 1968 and 1982, a total of twenty-five antitrust suits were brought against IBM—none of them substantiated in the courts. A long and bitter suit brought by Memorex was fought all the way to the Supreme Court before being settled in IBM's favor in 1981, and early the following year the Justice Department dropped as "without merit" an antitrust action that had tied up armies of lawyers on both sides for nearly a decade. By coincidence, IBM's reorganization was going on when the Justice Department dropped its action, and some staffers at IBM sniffed a cause-and-effect relationship. Not so, said Opel—the reorganization had been in progress long before, and had already been completed within the factories by the autumn of 1981.

In 1982, the FBI had arrested a number of employees of well-known Japanese firms for stealing IBM proprietary data. Those in the know in Silicon Valley were saying that the trap had really been set for Russian spies. Opel commented that the affair concerned design information rather than basic patents—information that would give a competitor several years' lead time to pirate a new computer design before IBM could bring it out. On such issues, said Opel, the company would take a hard line.

The European Economic Community (EEC) was pushing IBM to release such design information in advance of the first shipment of each new product, under a program called "Interface." Opel refused. But on questions of basic patents, he pointed out that IBM, because of its visibility and its wealth, has far more to lose than to gain from lawsuits. Company policy is to seek royalty-free cross-licensing, in order to be free of the risk of unintentional patent infringement. In that process, IBM and a counterpart firm evaluate each other's patent portfolios, agree to a royalty-free exchange and reevaluate at intervals to be sure the exchange is still fair. On the protection of basic ideas, Opel's view is that the company should err, if at all, in the direction of letting creative people interact, rather than erecting internal barriers to the flow of information.

Growth at least as fast as its competition is IBM's goal. In 1981, IBM's overall growth rate of 17 percent exceeded the industry average. But in that year the dollar recovered in value against foreign currencies as much as it had lost in more than a decade. IBM, which does nearly half its business overseas, saw $2 billion in gross receipts and $600 million in profit wiped out by currency conversion. In 1982, Opel felt that the company was positioned well for subsequent growth:

> This year we're up in shipments. We've borrowed at very good rates and made the big investments in our factories, so we have reserve capacity. We've developed the new product lines like thin-film disks and the Clark boards, and we're ready for growth. We can be the lowest-cost producer.

269

I asked John Opel if he had any worries for IBM.

> I do—but it's a positive, creative kind of worry: that because of our size and complexity we'll find it impossible to keep the entrepreneurial spirit, the feeling that it's a small company where people can affect its course and affect their own destiny. The worry that we may get too bureaucratic. That's why we have the Special Business Units and Independent Business Units. They're like venture-capital operations—we give them funds to get them going, but they have to make it on their own. Don Estridge's Personal Computer unit is one; Robotics is another; Instruments and our Turnkey units are others.

Opel pointed out that IBM's European divisions are unionized, because in much of Western Europe unionization is a legal requirement. But as for IBM-U.S.:

> If the employees think they need unions, then we in management have failed. There has never been a need for people to organize at IBM to gain recognition or advocacy, on the presumption that if they didn't they'd be exploited. There just isn't an adversary relationship.

AMERICAN EDUCATION

IBM must compete in its recruiting for a painfully small pool of truly well-educated young people who graduate from America's colleges each year. As a result, even IBM occasionally draws a loser. Significantly, the quality of education in the United States is a chief concern for Lew Branscomb, the IBM executive who is free to choose his own creative targets. Branscomb knows that "Working smarter is a better productivity strategy than working longer and harder," but he measures the deterioration of mathematics and science instruction in our public schools by the facts that:

- Only 1 in 5 U.S. high school students ever takes a junior- or senior-level science course of any kind.
- Only half the students take a math course of that level.
- Only 1 in 7 of our high school students ever takes a course in chemistry, and only 1 in 14 takes a course in physics.

The results are destructive individually and socially, and they can be measured. There have been a substantial decline in patent activity by U.S. inventors, a decline in the percentage of American citizens among engineering graduate students and a miserably inadequate number of engineering graduates for our industries: per capita, only a third as many in America as in Japan, and only a sixth as many as in Germany. But at least there is a growing awareness of the problem. As Branscomb says, "Not since Sputnik in 1957 has the general public been so concerned about the quality of U.S. engineering and science."

One potential remedy has to be independent action by parents. "Computer camps" are springing up in attractive summer-vacation spots like Cape Cod, Lake Tahoe and London. As yet there are no "physics camps," but introducing children to personal computers at an early age may be the best possible way to give them an independent window into every form of knowledge, because—unlike high-quality teachers—interactive educational programs can be replicated for as many children as there are who need them. Already there are public schools in the United States well equipped with personal computers. The next step will be to give each child the exclusive use of a computer, both at school and at home. A computer with disk drives and a printer can be bought for half the price of the cheapest automobile, and during the mid-1980s a new generation of inexpensive portable computers will drop the cost much lower still, to far less than public-school districts spend per student today.

Branscomb sees the interactive computer revolutionizing teaching not only at the school but at the college level. He feels that interactive home video computers will provide students with better educations than the classroom teachers can, because of the personalized, one-on-one nature of the interactive mode.* In the future, with existing knowledge distributed in that way by electronic systems, universities will be free to concentrate on the generation of new knowledge.

PRODUCTIVITY AND GROWTH

The industrial robot is a prime example of America's technical leadership being brought to bear on the challenges of manufacturing productivity, quality and cost. Branscomb regrets that the nation which is using that technology most effectively is not the United States, but Japan. He sees Japan's first priority as low-cost, high-

---

* If that view, which I share, is correct, schoolteachers are likely to become guides to social interaction, as social skills and values cannot be taught one-on-one.

quality manufacturing, and its goal as market share. Quoting former Secretary of Commerce Peter Peterson, Branscomb notes that between 1870 and 1950, the annual productivity growth rate of the United States exceeded those of the United Kingdom, West Germany and Japan by only 0.6 to 0.8 percent. Yet that small difference, compounded over eighty years, made the United States the economic and political leader of the world.

Branscomb deplores our engineering schools' demotivation of student interest in manufacturability, process technology and production automation, and illustrates how different are the views on industry from the two sides of the Pacific:

> Too many Americans read John Galbraith and Daniel Bell and got carried away [by neglecting manufacturing in favor of service industries]. By contrast, in Japan engineers often prefer a career in manufacturing to one in R-and-D. As one Japanese engineer explained it to me, "If you were Japanese and your daughter came to you and said she wanted to get married to a promising young engineer, your first question would be whether he worked in the plant or in the lab. And your first thought would be: I hope it's the plant, because if so he probably has better prospects."

CONCLUSIONS

Americans must accept the facts that the successful factory of the future is one of the most intellectually exciting arenas in our society, and that we cannot prepare our youngsters for careers in production without top-quality education. We cannot compete unless we end the antagonisms that have divided government, management and labor. Individual citizens will have to carry out the necessary reforms. The U.S. Government has been for the last decades, and is likely to be in the foreseeable future, incapable of carrying out effectively any consistent, enduring program to improve American education or productivity. The best thing it can do is get out of the way. And the traditional union movement, based on a continuing adversarial relationship between labor and management, has outlived its usefulness just as thoroughly as have those "wizard conglomerators" of whom Japanese managers are so contemptuous. As Branscomb of IBM points out, there is no single villain: not business executives for being insufficiently visionary, not government for overregulating, not labor for being too impatient, not banks for

being too conservative, not business schools for being too theoretical, not engineering schools for demotivating students from careers
in manufacturing, not school boards for allowing a disastrous deterioration in math and science instruction—yet all of those, taken
together, have brought the United States to a dangerous position.

The root of IBM's leadership is its human orientation. It can well
be copied, particularly by new start-up firms in the United States
whose managers are relatively young and whose employees have
not yet joined unions. Branscomb's observations on the U.S.–Japanese economic conflict have that same flavor:

> A science-based, quality-production-oriented economy needs two
> principles: thrift and education; it requires a knowledge-rich, free
> society that invests and plans—not for the next quarter's earnings
> per share, but for our grandchildren's respect for our vision.

Indeed, as Branscomb appreciates, we should be grateful to the
Japanese for shocking us out of our complacency and self-indul-
gence. Without their example, we might have continued to slide. It
would be pointless to attempt a wholesale imitation of the customs
of the Japanese, which are rooted in their culture and don't travel
well; but there are "Japanese" principles which foster both productivity and human dignity: taking the long view and making people
productive by bringing out the best that is in them—in Branscomb's words, "through a focus on understanding technology in
depth, communicating effectively across the organization and mo-
bilizing a consensus on clearly defined long-term goals." Those
principles have been cornerstones of IBM's success through several
generations.

Successful American competition in this decade will not, by itself, solve the problem of how to muscle through the introduction
of revolutionary new technologies with great potential benefit to
society. But if we do not compete successfully now, we will be too
weak to tackle that deeper problem. For all our mistakes in the
recent past, Americans still possess the capability to work hard, to
learn from our competitors and to regain the leadership we once
possessed. We are an adaptive people, and that is one of our
strengths. The examples we need are already here: in education, an
increasing reliance on the motivation that only parents can provide,
and the ingenious adaptation of the interactive computer to individ-

ualized teaching. We have already developed in the United States the societal acceptance and the financial mechanisms that bring entrepreneurs and venture capital together for the seeding and nurture of high-technology start-up companies. Those companies can give our economy a unique and extraordinary vitality. And such major firms as Hewlett-Packard, Delta, DEC and IBM show that Americans are capable, in the right conditions, of working together as productively as any people in the world.

CONCLUSION

We set out to find the solution to the most serious problem faced by the United States in the last several decades: how to achieve new and enduring economic growth in a marketplace that is global in scale and fiercely competitive in nature. The alternative, to let our nation drift, is unacceptable to most Americans.

We found and sorted out the key elements of the solution. Some of those elements turned out to be broad and general, attitudes and methods that must change throughout our entire society if we are to succeed. They relate to the education of our young people and our adults, to the responsibilities of working people to their employers and of employers to all who work for their companies. They relate to fundamental attitudes of business managers toward short-term profits and long-term growth. And they relate in just the same way to the expectations of investors.

We found plenty of evidence that Americans are still pragmatic and adaptable, are learning essential lessons from Japan and Europe, and are therefore changing their attitudes and beginning to carry out the revolutionary changes in methods that our country needs. We found that new start-up companies, where most new jobs will be created, tend to have those attitudes and use those methods from their beginnings. We found also, reassuringly, that in our large corporations with the best historical track record of success, the winning combination of attitudes and methods was established long ago.

Extending those changes in attitudes and methods throughout our society, tough as it will be, is the easier part of the solution to our problem—easier because we are being led in that direction by examples of international competition which force themselves upon us with every foreign purchase we make, and by books, magazine articles and the events of the daily news. The harder part of the solution is finding and dealing with opportunities to open whole new industries with multihundred-billion-dollar potential. The most important of those opportunities are not generally perceived, so bringing them to public notice is a major purpose of this book.

Our success in establishing ourselves in those new markets at a substantial level, before our competitors do, will make all the difference between breaking even—the best we can hope to do in existing markets—and winning the international competition. To appreciate the new opportunities, we will certainly need the healthier attitude toward long-term growth that we are now beginning to develop. And to exploit those opportunities, we will certainly need to employ the methods that we are beginning to learn.

I believe that we can meet the challenge, and the rewards if we do will be immense. If we retain the humility to understand that we may learn profitably from others, both at home and abroad, and the knowledge that we must never stop learning, we can become again a nation of productive growth, with increased opportunity for all of our citizens.

SUGGESTED
READING

PART I

For new developments in Japan's competition with the United States and Europe, the best continuing sources are financial newspapers and magazines:

Business Week
Time
U.S. News & World Report
The Wall Street Journal
Fortune
Forbes

Occasional special issues in most of these publications combine articles from a number of different industrial fields, all focusing on Japan's competitive position.

PART II

In addition to the publications listed above, several magazines directed toward the general public are useful for those technology areas already recognized as affording major opportunities:

High Technology
Robotics Age
Science 84
Discover

In the field of personal and portable computers, several monthlies with large circulation provide the most current information:

Softalk (for Apple Computer equipment and programs)
PC (for IBM Personal Computer equipment and programs)

For genetic engineering, *Scientific American* is particularly strong. In the areas not yet recognized as affording major opportunities, the source material so far is in professional or limited-circulation journal articles and technical monographs:

For magnetic flight:
"Between Riding and Flying" (1982; 24 pp.), available free in English or German version from
>   Bundesministerium für Forschung und Technologie
>   Pressereferat
>   P.O. Box 20 07 06
>   5300 Bonn 2
>   WEST GERMANY

Additional updated information on the Emsland Test Facility and the Konsortium Magnetbahn Transrapid may be obtained in English from
>   Messerschmitt-Boelkow-Blohm G.m.b.H.
>   Advanced Transport Systems
>   P.O. Box 80 12 65
>   8000 München 80
>   WEST GERMANY

For developments in light-aircraft technology, including navigation and flight-control electronics, readily available sources are
AOPA Pilot (monthly); available with membership from
>   Aircraft Owners and Pilots Association
>   421 Aviation Way
>   Frederick, MD 21701

Business and Commercial Aviation (monthly)
Magazines available on newsstands, generally less technically oriented, are
Flying
Private Pilot

For developments in space technology, readily available sources are
Aviation Week & Space Technology
Satellite Week
Aeronautics and Astronautics; available with membership from

American Institute of Aeronautics and Astronautics
    EDP Department
    1290 Avenue of the Americas
    New York, NY 10019
  SSI Update (bimonthly); available with $15 membership from
    Space Studies Institute
    P.O. Box 82
    Princeton, NJ 08540
Japanese progress in space is covered by publications in English from three main sources, the Keidanren (Federation of Economic Organizations), the Science and Technology Agency and the Institute of Space and Aeronautical Sciences. See especially:
  Space in Japan; published annually by
    Space Activities Promotion Council
    Keidanren
    1-9-4 Ohtemachi
    Chiyoda-Ku
    Tokyo
    JAPAN
  Proceedings of the First ISAS Space Energy Symposium (1982)
    Institute of Space and Aeronautical Sciences
    4-6-1 Komaba
    Meguro-Ku
    Tokyo 153
    JAPAN
and publications of
    International Space Affairs Division
    Research Coordination Bureau
    Science and Technology Agency
    2-2-1 Kasumigaseki
    Chiyoda-Ku
    Tokyo
    JAPAN
  Information on the Ariane rocket vehicle and the Arianespace launch company is available from
    Arianespace
    1, rue Soljenitsyne
    91000 Évry
    FRANCE

PART III

The general publications listed above for Part I are all relevant to both chapters of Part III. In addition, the current status of financing of start-up companies is focused on by the monthly *Venture*.

The large companies described in the final chapter are public. Current information on them is best obtained from their annual reports and 10K forms. IBM is particularly careful to distinguish its publicly circulated publications from others, generally published with equal professionalism and glossiness, that are restricted to its own employees. An excellent IBM publication that is openly distributed is the bimonthly ($4)

IBM Journal of Research and Development

Its 25th Anniversary Issue (September 1981, $6) is especially useful. Issues of this journal can be obtained from

IBM Journal of Research and Technology
Circulation Department
Armonk, NY 10504

INDEX